Technology Transfer to Latin American Countries

Hannig Núñez analyzes the processes behind technology transfers at a state-decision-making level in Latin America. She challenges the conventional notion that the United States and China hold a dominant technological presence over the region, highlighting the increasing influence from both middle powers and regional actors.

This book builds on existing theory and case studies to assess the relevance of economic incentives, geopolitical rivalries, and value-driven considerations in the outcomes of technology transfer in different scenarios. It further explores the notion of a new "Cold War" between China and the United States and examines how these superpowers leverage technology transfers to extend their influence but ultimately fall short due to growing competition from previously overlooked players. In closely examining these dynamics, Hannig Núñez demonstrates how technology transfer is not solely an economic process but a significant geopolitical tool that influences international order, national sovereignty, and regional integration.

An invaluable resource for students, academics, and researchers interested in the intersection between technology, cybersecurity, and international politics.

Sascha Hannig Núñez is an executive secretary at CAD Chile and a researcher at the Center for Global Governance Research, Hitotsubashi University, Japan. An international analyst with an interest in technology, she consults for global organizations and has experience as a financial reporter.

Routledge Advances in International Relations and Global Politics

The Foreign Policy of Irregular Migration Governance
State Security and Migrants' Insecurity in Italy and Australia
Gabriele Abbondanza

Mexico-China Relations
Cultural Encounters in a Global Age
Francisco Antonio-Alfonso

Power and Influence in the Pacific Islands
Understanding Statecraftiness
Edited by Joanne Wallis, Henrietta McNeill, Michael Rose and Alan Tidwell

Changing Abortion Laws in Mexico Through Advocacy and Human Rights
When Federalism Came to Life
Clara Eugeina Franco Yáñez

Europeanisation and Eurasianism in Turkey and Ukraine
Balancing between Integration and Autonomy
Lucie Tungul

Human Rights, Impunity and Anti-Press Violence
Journalists' Strategies for Countering Unpunished Attacks
Tamsin S. Mitchell

Technology Transfer to Latin American Countries
Drifting Away from the United States and China?
Sascha Hannig Núñez

For information about the series: https://www.routledge.com/Routledge-sAdvances-in-International-Relations-and-Global-Politics/book-series/IRGP

Technology Transfer to Latin American Countries

Drifting Away from the United States and China?

Sascha Hannig Núñez

First published 2025
by Routledge
4 Park Square, Milton Park, Abingdon, Oxon OX14 4RN

and by Routledge
605 Third Avenue, New York, NY 10158

Routledge is an imprint of the Taylor & Francis Group, an informa business

British Library Cataloguing-in-Publication Data
A catalogue record for this book is available from the British Library

ISBN: 9781032788326 (hbk)
ISBN: 9781032788340 (pbk)
ISBN: 9781003489450 (ebk)

DOI: 10.4324/9781003489450

Typeset in Times New Roman
by codeMantra

Contents

Preface

Societies have long relied on technology to make life easier, safer, and longer. At the same, more advanced technology can give a group or state a comparative advantage over others, not only in terms of military capabilities but also economic and cultural, making the production and acquisition factor an essential part of power. In a world where almost 70% of humanity is connected to the internet, sovereignty and value-related questions need to be addressed by decision-makers. Some countries are transparent technology producers, and others adapt and acquire technology to maintain their competitiveness and take advantage of better systems, cost reductions, and enhanced capabilities provided by innovation.

In this book—the result of a three-year research project—I discuss a particular aspect of the relationship between technology and international relations—the transfer of innovation and decision-making mechanisms in Latin America amid the increasing rivalry between the United States and China. This endeavor required extensive research, consultation with primary sources, and a dive into narratives spreading across Latin America. It also required the comparison of existing narrative and theoretical frameworks that have attempted to analyze the region as a political bloc in the context of the US-China rivalry and in the matter of technology transfer. In that context, I expect the cases presented in this book to also reflect the commonalities of Latin American states, as well as the nuanced and sometimes dramatic differences in governance and political decision-making, so relevant to understand the outputs of these processes.

I sincerely believe that this work will contribute to the theory surrounding technology transfers, the understanding of Latin America, and the current discussion on the idea of a modern Cold War.

Acknowledgments

I sincerely thank the individuals and institutions whose support made this work possible. First, I must express my gratitude to my supervisor, Dr. Maiko Ichihara, who guided me through structuring, writing, and strengthening the theoretical framework of this book. Also, I need to thank her for her help in tackling the limitations of this work. On that same line, I would like to thank Dr. Rafael Rincón, who helped me visualize the interactions between technology and society across history from a multidisciplinary perspective. There is a third person whose name I cannot openly disclose but whose insight on policymaking in Chile and help to contact sources in the public sector directly were instrumental in the success of this project.

I welcome the institutional support of the Graduate School of Law of Hitotsubashi University, Japan, and the Japan International Cooperation Agency (JICA). Both institutions helped fund my research and have been instrumental in successfully delivering this book and disseminating its findings. In addition, I would like to thank my colleagues at the Graduate School of Law, Hitotsubashi University, for their insightful comments on my draft and Chrystopher Kim for his thorough review of this book's conclusions and recommendations. Also, to Benjamin Sando for his review of the first chapter. Along the same line, I'm grateful to Pablo Zeballos and IBI Consultants for providing the photographs of their field work in Argentina, and their permission for these to be used in this work.

Of course, I need to thank the Routledge editorial team, particularly Clarissa Lim and Khadijah Ebrahim, for their continuous support in completing this work and for their guidance in form and structure.

Finally, I would like to thank all the stakeholders and experts I interviewed for this work, the general information of whom is presented in the bibliographic sections of this book. Their views of this issue profoundly helped me map the characteristics of technology transfers, the role of the private sector, and the increased challenges that geopolitics pose to emerging economies.

Part 1

Theory, Technology Transfers, and the Latin American Geopolitical Scenario

1 Introduction

Technology Transfers to Latin America amid Increasing Geopolitical Tensions

In 1939, the Spanish philosopher Ortega y Gasset asserted that the existence of humanity is inextricably intertwined with the knowledge of technology or, in his words, that "without knowledge over technique [a broad notion of technology], humans would not exist."[1] He intended to emphasize the need for the social sciences and philosophy to explore material change and the logic by which technology operates and affects hierarchy, power structures, governance systems, social relations, and culture.[2] In other words, innovation plays a central role not only in the progress of societies but also in how different individuals, groups, and—I argue—states compete to be at the forefront of development. Following this line of thought, Byron Reese links human historical progress in terms of the technologies they were able to master.[3] In Ferguson's *Civilization: The West and the Rest*, he directly identifies how philosophical currents, policies, and institutions shape the competitiveness of societies—and by extension, states.[4] Coincidentally, Boehme-Neßler dedicates a considerable share of his argument to explaining the interconnection between technology and culture and their impact on social development.[5] In contrast—and related to the Latin American context—Marxist authors Eduardo Galeano and Belfrage also connect development to policies and access to technology.[6] However, their theory is more deterministic than the two previous examples, and I argue that it should be revised, something I tackle in the following chapters.

Recent examples concerning the international community evidence this relationship. Notably, only about two-thirds of the world's population has direct and constant access to the global internet network,[7] and groups connected to the internet also gain access to information and knowledge that can give them an advantage over those who are not connected. The same is true for those groups with greater access to technology—which can mean higher economic outputs. At the same time, the distribution of medical supplies—including vaccines—during the COVID-19 pandemic exposed significant inequalities in the production and distribution of medical care across borders. This is a clear example of how geopolitical conditions are shaping the adoption, transfer, and deployment of innovation—with economic security as a crucial point of discussion on the limits of trade under increasing tensions in the international theater.

DOI: 10.4324/9781003489450-2

Generally, scholars will share the notion that the dynamics of international relations (IRs) can be related to a state's access to technology,[8] and they do this by understanding it as part of political processes or by incorporating it into traditional notions of **power**. This means, for example, adding innovation outputs to a state's military capabilities or enabling economic pressure over weaker or poorer nations. However, the literature on this relationship is limited.[9] From a theoretical perspective, Morgenthau's work creates distinctions between material power—in the form of military capabilities—and political power,[10] which is essential to establish the range in which technology affects a state's dynamics, especially in the digital age. Gilpin pointed out that technology could affect a state's position and influence.[11] Nye has also touched upon the inclusion of technology in his definition of state power, considering material and immaterial measurements.[12] In that sense, his conceptualization of soft power considers the cultural influence and reputation of technologies—and how these affect power—and its combination in different forms of hard power as smart power.[13] Following this logic, Weiss has highlighted the role of technology not only in the traditional material notions of material power but also in the perceptions and societal responses to innovation.[14] In addition, when measuring and comparing state power, some indexes, such as the Global Power Index (GPI), use technology as a variable to compare power between different states.[15] However, these indexes are often flawed in measuring certain aspects of non-tangible power.[16]

As for the framework connecting power, technology, and IRs, the literature has been approached from critical deontological perspectives. McCarthy tackles this question from a constructivist approach and points to the need to understand technology in terms of its "social power," with which changes can be triggered in both domestic and international contexts.[17] Skolnikoff, on the other hand, approaches this interaction—of global affairs and technology—regarding its effects on sovereignty and the potential for societal change[18] and clarifies the distinctions between science, technology, and the ambiguity of these terminologies.[19] More recently, Eriksson and Newlove-Eriksson have reviewed the existing bibliography on the interactions between technology and IRs, assessing the need to determine a more integral point of interconnection between politics and technology—and argue that this transcends traditional theoretical notions. At the same time, they open a crucial question on technology's political nature and innovation objectives.[20]

It is also necessary to explore new challenges in interpreting the nature of technology and its relevance to power. These challenges include the active role of the private sector in producing these tools. Considering that IRs studies have traditionally put states at the center of analysis,[21] exploring the dynamics of private actors in the research, production, and diffusion of technology constitutes a counterintuitive but necessary analysis point in this work. This is something that authors Eriksson and Giacomello emphasize. In their argument, while IRs studies have mostly remained state-centric, it is impossible to deny the pivotal role that the private sector and the IT community—including non-profit organizations—play in game-changing innovation.[22] They explain the gaps in the existing theories regarding the inclusion of technology in IRs analysis. For them—and this is reemphasized in the most recent Eriksson and Newlove-Eriksson argument—realism, liberalism, and constructivism are not enough to explain the shift of power in new technologies in the context of a globalized world.[23]

In other words, analyzing the effects of technology in IRs requires a framework that incorporates the agency and role of technology producers, even if these are not directly associated with state power. Of course, some states have a more significant influence on their private sectors, which could give them advantages in direct competition with other states and in the weaponization of technology for military, surveillance, or cultural influence purposes. Because of this interaction of states and technology, concepts such as Joseph Nye's soft power and Edmund Gullion's public diplomacy gain theoretical importance.[24]

In this book, the neoclassical realist framework can provide a valuable answer to the questions of technology, power, and state interactions regarding technology transfer across borders. This approach to IRs theory integrates systemic and unit-level variables[25] and can incorporate domestic and international factors in explaining foreign policy decisions.[26] Ripsman finds this approach convenient for understanding the nuances of a given context because it allows for revisions and considers fundamental principles from liberalism and constructivism, as the interests of domestic actors are not always rational or pessimistic.[27] However, following Gourevitch's argument, neoclassical realists do not directly claim that the international economic system or international institutions will impact policymaking,[28] except when following said norms increases their relative position in the international system. In this context, Robert Putnam's "two-level games" framework is also helpful in describing the increasing importance of these private actors and operationalizing how domestic and international factors can impact policy beyond classical state interactions.[29]

Acquiring and Transferring Foreign Technology between States

In coherence with the existing IRs theory, there are several pragmatic and deontological reasons for states to emphasize the creation and production of technology while striving to disseminate their innovations globally. This drive is focused not only on economic gains but also, somewhat contradictorily, on positively impacting society and increasing their relative power in a highly competitive international arena.

Since investing in technology, research, and innovation is highly costly and doesn't always guarantee success, some states prefer replicating, harmonizing, and transferring technology abroad. The process of acquiring or diffusion technology from one state to another is generally called "technology transfer." It is of great interest to both innovators and those who receive a particular technological tool. States can gain power by harmonizing their technologies across borders since these often are incompatible and require the receptive country to choose a specific provider. At the same time, countries can use existing technology with little risk—in terms of research and development investments—and skip developmental phases that industrialized countries had to go through.[30]

At the same time, developing states could be at an advantage compared to industrialized states because they can access existing knowledge, tools, and policies that took rich countries centuries and large amounts of resources with high risk. Gerschenkron called this "backwardness advantage" in his 1962 work: "Economic Backwardness in Historical Perspective."[31] In this sense, while existing powers see the potential for diffusing their technology internationally and developing countries can access complex technologies at an affordable cost, emerging technology producers are also at an advantage, as they can quickly catch up and become competitors to the existing industrialized powers. This is achieved by the replication of technology and the long-term implementation of policies that impact the development and diffusion of innovation.[32] In cases like Korea or Taiwan—and Japan, much earlier—industrialization and technology production have resulted in outstanding economic growth and an international reputation in the field. Meanwhile, China has also shifted to a technology-focused strategy with explicit intentions to internationalize.

However, with growing concerns about the use of and access to data and general skepticism about the interplay of values and technology production, geopolitical debates are finding their way into decision-making as states decide which vendors to choose when implementing innovations that have a large social impact, such as telecommunications, surveillance, or private data management.

Moreover, advanced technologies such as generative artificial intelligence (AI) and forms of hybrid warfare involving disinformation, hacking, and general disruption of telecommunications are rapidly developing. The latter is important internationally, as many governments see telecommunications as key to their domestic development[33] and an area of innovation that is quite susceptible to the security threats posed by data transmission over networks.[34]

The Geopolitical Rivalry between the United States and China

A prevalent viewpoint is that at the heart of the transfer of technology question lies the rivalry competition between the United States (US) and the People's Republic of China (PRC).[35] Both states maintain intensive technological production, and their interests and influence extend beyond their immediate borders. The so-called trade war[36] between them has been widely covered by scholars who, for example, examine the diversification of the US economy away from Chinese manufacturing[37] or the direct consequences of this economic rivalry.[38] Some predicted an increase in competition as soon as China entered the world market through the World Trade Organization (WTO) and sought to expand its international presence.[39] Regarding technology, I found that a considerable amount of research is focused on mapping the roles of the PRC and the US in the diffusion of sensitive innovations around the world.[40] There is evidence, however, that technology transfers in the context of national and international policymaking do not only involve these two powers. Other increasingly relevant actors have addressed security issues between private and public data. In countries such as Australia and some European nations, mechanisms such as tightened screening processes for security reasons have been in place for decades.[41] As a result, some companies worldwide are restricted from participating due to noncompliance with global standards. This leads to discussions on harmonization processes and a more comprehensive discourse regarding values and ideas behind the decision-making process. In Latin America, there have also been some specific cases of transfer limitations.[42]

Consequently, the outcomes of highly relevant public tenders related to technology transfer do not benefit either significant power to the extent that might be expected given the publicity and intense debate about Beijing's competition with Washington. In this book, I explain how the focus on US–China tensions diverts attention from a broader international phenomenon, and although it is essential to acknowledge the presence and influence of these two actors, focusing only on them can lead to bias. For example, in the case of 5G networks, research sometimes is centered exclusively on the abovementioned dichotomy between Washington and Beijing. Still, only a few American vendors appear in public tenders or compete to challenge Huawei in developing countries. In contrast, actors such as Nokia (Finland) and Ericsson (Sweden), as well as development organizations such as the Inter-American Development Bank (IADB) and the OECD, are increasingly pushing to compete "outside" the rivalry or bipolar situation and are less discussed in the public debate.

Moreover, when I analyzed the state of public opinion and media behavior related to the debate over 5G networks for this work, I found that private actors from origins outside the US and China are rarely mentioned compared to the more controversial Chinese companies. This raises the question of whether concern about the spread of 5G networks in the region is due to the global discussion or to grassroots movements in telecommunications. There is an urgent need to explore this narrative and contrast it with cases of geopolitical concerns and I tackle that gap within this book.

It must be clarified that the absence of American or Chinese private actors in a technology acquisition process does not always mean these countries lack influence in international innovation transfer dynamics. The concept of a *new* or *second* Cold War is also becoming prevalent in academic discussions surrounding the transferring of technologies and screening of providers. Under this notion, the decisions made by states are not aligned with a specific country but rather with general values and interests that are easily identified in debates about private data or freedom on the internet. However, there is no consensus on whether using that term is appropriate to reflect current divisions between the democratic world that promotes the International Liberal Order (ILO) and an ongoing alliance between anti-liberal nations and authoritarian governments seeking to change international norms. A line of critique points out that while these partnerships are convenient, globalization has made it almost impossible to reach the same level of economic and political division witnessed between the Soviet Union and the US in the past century. A cold war would imply the existence of formal and informal alliances and other actors choosing a side based on value affinity or convenience.[43] During the Cold War, these alliances reflected a set of political and economic system features, consequently shaping technological achievements' success and failure.[44]

Although the current situation has particular characteristics, this book explores the notion of a "new Cold War" by analyzing cases that could support or debate the idea of technology transfer in the region. The results of the case study section of this book generally show that Latin American states are aware of the rivalry between the US and China. The region's history and the political values of different governments are also discussed to explain the outcome of specific decision-making processes, the role of reputation, and the association of certain private actors with their countries of origin.

Latin American Countries: Early Adaptors but Not Innovators

The contribution of this book is broad and focuses on technology transfer to Latin America in this complex geopolitical landscape. Much of the book is devoted to South American cases because the subcontinent is home to nearly half the Americas population and has varying degrees of success in governance and economic systems. The countries of Latin America are not generally thought of as innovators in the strict sense of the term,[45] with some exceptions in scientific discoveries such as the typhoid vaccine,[46] or software services that complement the needs of their citizens, with examples such as Mercadolibre in Argentina or Cornershop in Chile. However, developing countries tend to adopt emerging technologies at the national level, resulting in a higher percentage of connected citizens compared to other developing regions. Trends point to Argentina, Brazil, and Chile as the current leaders of adaptation in the region. At the same time, Uruguay and Colombia have updated privacy regulations and legal frameworks related to technology, mirroring the European General Data Protection Regulation (GDPR).[47]

According to internet World Stats, by the end of 2023, 84.6% of the population in South America and 80.5% in Latin America and the Caribbean was connected to the internet, compared to 67.9% globally. The least connected countries in South America are Suriname, with 71.9% of the population connected, Guyana, with 72.5%, and Bolivia, with 73.9%. All these internet penetration rates are above the world average despite the objective dependence on foreign technology providers for said development. The most connected countries are Chile with 97.2%, Uruguay with 93.2%, and Argentina with 91.1%. Two outliers—both overseas territories of European nations—deviate from the regional trend. The first is French Guiana, with a 52.2% penetration rate; the second is the Falkland Islands, with 3,653 inhabitants and an outstanding 98.5% penetration rate. The statistics from Central America are staggering. Countries like Honduras have internet penetration rates as low as 37.6%. The regional average for internet penetration is 61.9%, with Costa Rica reaching the highest rate at 85.9% and Panama following it at 68.6%.[48] Finally, Mexico, which has an internet penetration rate of 66.5%, is sometimes considered a Central American country and sometimes a North American country. Either way, Mexico is a crucial case because of its influence in the region through companies such as Claro and as a reference point for policymaking in Latin America.

It is evident that the distinctions between South and Central America are significant and require further examination. One could argue that this context presents an opportunity for certain companies to play a pivotal role in developing the telecommunications infrastructure for Central America's approximately 100 million individuals. Local sources assert that the distribution of internet access is not uniform, citing disparities in quality, connectivity, and economic factors as key determinants. This leads me to two crucial points that this book must establish early. First, the financial conditions of a country are almost certainly linked to internet access, which must be considered when drawing conclusions. Secondly, internal inequality may be the cause of a set of attitudes across the public. Despite a high percentage of internet access, South American countries tend to be skeptical of science and technology,[49] and the most recent iteration of the World Values Survey offers insight into this skepticism. When asked to evaluate the impact of science and technology on the world's well-being on a scale of 1–10, the global sample demonstrated a clear preference for the upper end of the scale, with 21.8% selecting the maximum score for "better off."

In contrast, only 7.2% of Argentinians, 12% of Bolivians, 5.3% of Chileans, 14.5% of Ecuadorians, 14.3% of Guatemalans, 18.5% of Mexicans, and 5.8% of Peruvians responded with that amount of certainty in support of technology. The only exceptions within the sample were Brazil (24.8%), Mexico (18.5%), and Uruguay (19.9%), where most respondents expressed certainty in their support for this statement.[50] Future chapters in this book will address values and societal responses to this issue. These will explore skepticism toward technology and surveillance technology, reflecting how their citizens will react to foreign innovation and their understanding of the social benefits and barriers accompanying policy responses.

In this book, I posit that Latin America's technological adaptation is significant in IRs studies, given the region's unique characteristics and relevant powers' interest in these countries. However, research in this area is still limited or fragmented. Research suggests that the region has been the subject of competition between the US and the PRC, both in terms of traditional investment and technology diffusion. Organizations concerned with regional development and poverty alleviation have also contributed to this perception. For example, Chinese investment has increased significantly. It has been referred to by the United Nations Economic Commission for Latin America and the Caribbean (ECLAC) as a "civilizing connectivity proposal for shared prosperity."[51] In the same vein, research has revealed China's active and overt attempt to increase its regional presence, mainly through economic statecraft and public/economic diplomacy.[52]

A notable proportion of studies concentrate on historical analysis or political interpretations of the issue. In some instances, these appear to deliberately disregard the involvement of other actors or integrate alternative data sources. For example, during my research for literature relevant to this book, I came across an article, frequently cited and downloaded, from Colombo et al. addressing the role of 5G in the technological competition between major powers. It criticizes the US for blocking Chinese technology in Brazil and Argentina, citing recollections of tweets, news articles, and correlations to support its claim of interference in the region.[53]

The issue that arises when attempting to comprehend this phenomenon is the assumption that the US competes with China under the same conditions and policy environment, exerting influence on US companies to be selected as providers or directly controlling their actions. Firstly, implementing a specific technology frequently depends on the collaboration of multiple providers, often from disparate countries, industries, and final owners. As discussed in the forthcoming chapter on the popularized 5G debate, the US and China are not engaged in direct competition in most public bids for 5G infrastructure. In some cases, private or semi-private companies select a specific provider, the identity of which is not disclosed to the public until the commencement of work. Some companies utilize a combination of Chinese and US technology, contingent upon the specific service in question. Secondly, the structure of the Chinese Communist Party and its capacity to direct a diverse range of public and private enterprises make Beijing a distinctive actor within the region. The US has expressed concerns about Chinese telecommunications providers. Still, it lacks the ability to compel American companies to invest or participate in public tenders or bids in South American nations without incentives. As a result, competition assumes a narrative and diplomatic form rather than direct actions.

Nevertheless, while it is essential to consider the interests of these two countries, it is also necessary to acknowledge the efforts of companies from other territories seeking to participate in the region. Japanese firms have demonstrated a heightened interest in local security contracts, while a recognizable French company has established a dominant presence in providing identification-related services.

Additionally, countries such as Israel, Sweden, and South Korea have emerged as crucial sources of cybersecurity-related sensitive technologies. In addition, Latin American governments have selected alternative options to the anticipated US or Chinese ones, including European, South Korean, or Japanese actors' technologies for innovations such as 5G networks,[54] facial recognition software,[55] fiber optics standards, and even national databases.[56]

Notably, some of these countries are confronted with governance-threatening issues. It is evident that cybersecurity systems are not always updated and cannot contain every potential attack that regional administrations may face in an increasingly digitalized world. As previously stated, some governments are amending their legislation to align with the European Union's GDPR framework, thereby fostering greater harmonization across disparate legal jurisdictions.[57]

The existence of competing companies and the disparity of levels of competition are not the only factors that impede the development of a theory of exclusive competition between the US and the PRC. Citizens across some countries in Latin America also view both nations with skepticism, so public opinion does not always support the idea of these two actors having direct or indirect control of critical technologies and infrastructure.

South American countries provide an intriguing case study for examining how technological evolution at the national level can and should be analyzed from an IRs perspective. Concurrently, it demonstrates the significance of theoretical frameworks in contextualizing the phenomena under investigation in this book. This is because the region encompasses a range of democratic systems, from Venezuela to Uruguay, as well as contrasting economic systems. As a result of the shared cultural traits and historical factors among much of the bloc, case studies can be compared based on their similarities. Consequently, the findings can be generalized to other studies on the behavior of the US and China in Latin America.

Considering the aforementioned factors, this book makes a significant contribution by providing a comprehensive and rare approach to the analysis of technology transfers in Latin America. It offers a nuanced understanding of the wider geopolitical competition, while also acknowledging the role of other middle powers and regional leaders. This demonstrates the complexities inherent in the region's decision-making process and the considerable variation in Latin American states' approaches to technology. In this book, I examine how certain governments have sought to establish themselves as regional leaders and innovation hubs by eschewing the US and China, while others have been compelled to navigate the complexities of maintaining close relations with both powers.

How Values and Geopolitics Interact in the Region

To write this book, I interviewed sources from the private sector, international organizations, academia, and various public officials and political actors.[58] Their testimonies are reflected throughout the arguments, and their individual narratives are also detailed in some chapters. In general terms, these sources confirmed the dominance of a "US vs. PRC" discussion not only in the research community but also in policy circles.[59] The general assumption is that geopolitical tensions affect private decision-making and that this spills over into domestic and international discussions.[60] As countries seek to balance their relations with both Beijing and Washington, the alternatives for these countries become more attractive. Nevertheless, some argue that "not choosing" Chinese suppliers is already a form of compliance in favor of the US,[61] as the debate may focus more on where countries will position themselves in the eventual consolidation of a "new 'Cold War'." In this scenario, technological competition is seen as part of a global conflict involving middle powers and developing countries.[62]

This chain of influence and cultural considerations is linked to the concept of soft power and its role in influencing the policy outcome within the country under study. Indeed, the reputation of a provider was frequently identified by interviewees as a relevant factor in predicting the success of a public tender, even in the absence of formal screening. In contrast, the media content analysis conducted for this study revealed that the term "reputation" is not utilized by media outlets when examining matters pertaining to technology transfers. The analysis of 5G deployment in the region yielded the primary finding that technical and geopolitical keywords were the most prevalent in the review. As with the general geopolitical influence theory, the media appears to reflect the tensions between these two main powers when discussing, in particular, the implementation of 5G technology and the involvement of Huawei. In light of the aforementioned mentions, it is noteworthy that other pertinent companies, such as Nokia and Ericsson, are referenced with lesser frequency.

Technology and IRs: Balancing the State, Private Sector, and the End-User

This academic work dedicates several chapters to establishing a concise nexus between innovation and IRs theory. This serves to illuminate the significance and particularity of the cases under study and to further extend the relevance of this book to new and emerging phenomena that are likely to arise in the near future.

The relationship between state-level decisions, technology, and society is contingent upon a multitude of factors, including the domestic circumstances of the decision-makers themselves. In essence, the implementation of a legal framework to regulate or integrate technologies into societal structures has a profound impact at the national level. When these decisions are discussed across multiple countries, the complexity and impact of such decisions increase. The relationship between nation-states and the diffusion of innovation has been the subject of analysis by IRs theorists. These analyses have employed realist, liberal, or constructivist frameworks to draw parallels between traditional notions of power and the influence of technology in the accumulation of power. Considering these developments, it is reasonable to conclude that technological innovation has the potential to challenge and affect international dynamics in a multitude of ways. The advent of the radio, telephone, and nuclear power during the 20th century are illustrative examples of innovations that have not only transformed people's lifestyles but also have had a profound impact on IRs. Since the 1980s, the introduction of the internet and a more sophisticated form of AI, espionage, surveillance, and cyber threats have continued to present challenges to this day.[63] The relationship between technology and global affairs contains two elements that make analysis more challenging. The first is the crucial part that the private sector plays in development, decision-making, and investment outputs.[64] Due to the frequent and rapid changes in the information technology sector and the emergence of new threats from digital interactions, states often need more capabilities to effectively manage and address these issues. This is primarily due to the complex and rapidly evolving nature of the sector, which frequently operates in a "grey zone," making it difficult for states to keep up with the latest developments and effectively respond to emerging threats.[65]

Drawing on more than a century of experience in technology transfer and diffusion, countries with both democratic and authoritarian regimes agree that norms are needed to engage with these issues from an IRs perspective.[66] At the same time, both democracies and autocracies support the idea of collaborating and dealing with "like-minded countries"[67] and pushing for certain values or limitations to freedom within international conventions.[68]

In addition to the normative standards, studies have also been conducted on security-related issues that affect the flow and infrastructure of information.[69] The transfer of a particular technology or standard involves several factors, such as international agreements, expanding privacy laws, national critical infrastructure, data protection regulations, and the need for national or regional harmonization to achieve competitive advantage and cost reduction.[70] Because of these factors, technology harmonization, a term used to describe the existence of compatible standards to easily transfer and adapt systems that require expensive infrastructure, has been approached from several angles, including international public policy and commercial strategies in product manufacturing, including quality and specifications.[71]

The interaction of IRs with technology has been analyzed under most theoretical frameworks. Realism is an obvious candidate when discussing the use of innovation in warfare and its security implications; liberalism when discussing trade or international institutions; or constructivism when analyzing the social and psychological aspects of technology implementation.[72] However, due to the multidimensional nature of technological creation and implementation, all IRs theories face methodological challenges in explaining key aspects of these interactions.

Technology transfer is no exception when facing these limitations. Firstly, this is a result of the multi-dynamic nature of actors, regulations, and elements involved. When a multinational company decides to compete in a foreign public tender, it brings influence from its origin country or countries, such as culture, regulations, and reputation. Perhaps for this reason, neoclassical realism is more often mentioned in research as a framework to overcome these epistemological limitations while keeping security and traditional conceptions of power at the center of the discussion. In other words, this framework can incorporate domestic factors and decision-making elements to understand the international behavior of states and how international structures are shaped by issues that are not normally considered in a purely realist view. At the same time, this perspective understands that foreign trends affect domestic policies.[73] This last point is important because, in a globalized world, nations are less and less able to remain isolated from the transfer of innovation—and the conflicts that surround it. In other words, countries cannot simply choose to remain outsiders to these trends and issues if they expect to remain relevant in the international sphere.

To better explain this proposed relationship, I dedicate some chapters of this book to the theoretical relationship between the phenomenon of technology and international politics. I examine the existing literature and provide case studies to support this proposal. The intention is not to yield a solution to the debate on approaching practical issues with different schools of thought and paradigms but to highlight the research gap that exists and continues to grow with innovation and development.

A Philosophical Debate on Technology Adaptation

Although not a primary goal for the theoretical contribution of this book. I present a historical and philosophical framework that I believe is the best for understanding the cultural impact of technology beyond practical examples that are too focused on the economic consequences of technology and innovation. From the discovery and mastery of fire some 100,000 years ago to the remarkable technological advances of the present day, the ability to devise solutions has played a significant role in shaping human social and political interactions. As Boehme-Neßler explains it: "The influence of the state, culture, and society on technical development can be frequently observed in cultural and technical history (…). So social, political, economic, and cultural parameters all have a major influence on the development of technology."[74]

This science-based or cross-disciplinary philosophy can be challenging, especially in a society that often separates science and social studies. Politics involves legal frameworks, interpersonal relationships, and analytical reasoning, while science strives to improve living conditions and deliver new inventions. However, science also has the potential to provoke danger through the development of weapons and dehumanize society with its discoveries. For this reason, tribes, empires, and, more recently, countries have pursued technological advances that put them one step ahead of their competitors. Technology has thus also shaped historical outcomes. Technological improvements in irrigation and water supply to cities were key to the urban development of the Roman Empire for centuries,[75] and are also seen as a potential—though debated—reason for Rome's decline.[76] The Meiji period in Japan is widely regarded as a time when the need for rapid modernization was recognized by its leadership. The actions taken by the government during this period reflect the importance of keeping up with the rest of the world in terms of industrialization and military capabilities.[77]

In this sense, this book is based on the assumption that technologies, broadly defined, can affect political outcomes—from the development of local communities to interstate wars—and that, to some extent, they also have a say in the way countries act and react depending on their context.

It is important to note that not all discoveries have the potential to change the course of human development. However, major innovations, such as the telephone, electricity, or the internet, can trigger societal shifts that require changes in governance, as administrations seek both to provide a framework and regulate these breakthroughs to limit or spread them. A modern example of this is the use of AI and robotics.[78] In the same vein, the introduction of the innovations of the 4th Industrial Revolution into everyday life is changing societies around the world,[79] and some authors such as Fareed Zakaria argue that COVID-19 accelerated the digitization of life and, therefore, society's interaction with data and platform-based life.[80]

These changes become even more dramatic with automation and the introduction of AI to tasks that were thought to be exclusively human,[81] such as court decisions, medical examinations, or even artistic creation. AI can also be used by governments to measure, monitor, and control their populations. More abruptly, the incorporation of tools like Chat GPT, accessible deep fake tools, and other uses for these innovations seem to be something that governments and industry are seldom prepared to face.[82]

Establishing a Framework for Research on Technology Adoption in Latin America

This book consists of two parts. The first part is theoretical and begins by establishing the link between technology, development, and IRs. It also establishes a framework for the assumptions and baselines on which Latin American governments make decisions, along with the unique aspects of the influence of the US and China in the region. It concludes by characterizing the types of actors operating in the region alongside these two powers, drawing the line between the agency of private companies and their countries of origin. Throughout this section, other important factors, such as geopolitical struggles and important historical facts, will contribute to a better understanding of the current regional situation and establish general factors under which this book can serve as a reference to expand and complement this research.

Chapter 2 provides a philosophical and historical overview of this book's importance, linking the relationship between international dynamics and technology development and development in general. This chapter also helps illuminate the need for interdisciplinary research regarding innovation and technology development.

In Chapter 3, I present the state of Latin America's relationship with the US and the PRC. This chapter aims to frame the problem of geopolitical tensions and the influence of these two powers, as well as how this relationship has evolved—and to emphasize the importance of this book in presenting a perspective that has only been analyzed in isolated cases or by the media. Some historical issues, such as China's campaign for the de-recognition of Taiwan and its support of local dictatorships, are also addressed. In Chapter 4, I present a theoretical framework that is connected to the IRs literature and defend the use of neoclassical realism to organize the analysis of this book. I also introduce relevant concepts that explain technology transfer under economic and political power dynamics. Some of these are soft power and reputation factors, backwardness advantage, and harmonization. Chapter 5 focuses on three different reasoning mechanisms under which states would be likely to select a supplier under various competitive regimes or direct negotiation. These include decisions based on economic incentives, value-based considerations, and geopolitical pressures. These three reasoning schemes are considered later in the conclusions.

The second part of the book develops the theory by focusing on the specific cases that show how decision-making is influenced in Latin America. These cases reflect the idea of a region in which the US and China compete but also in which other actors actively participate and even win in technology transfer to the region. Therefore, in Chapter 6, I present a counterintuitive case study in the form of digital television standards in Latin America. This is an interesting case in which Japan, together with Brazil, played a dominant role—with several South American countries following suit—rather than opting for the American standard pushed by the US. It explores the successful role of middle powers in the diffusion of technology in the region—something that is important for understanding the outcomes of the following cases. In Chapter 7, I present several cases of private data use in Latin America, including the construction of a transoceanic cable between Asia and South America, Venezuela's identification card system supplied by a Chinese company, and competition for data center use in the region. In Chapter 8, I discuss another form of identification technology, digital surveillance tools, focusing on facial recognition. I also review the role of Chinese companies as suppliers, as well as the dominant role of Argentina and Brazil. In this chapter, I navigate the ethical discussion of security and privacy in the region, using these examples to convey the state of regulations and ethical considerations.

These chapters build on the theoretical framework of Chapter 9, in which I present the state of 5G spectrum adjudication in Latin America. This section of the book sheds light on the impact of political narratives, the indirect involvement of Chinese companies in the process, the influence of domestic phenomena on decision-making, and the role of other Asian, American, and European companies. It also discusses reputation and geopolitical pressures on the outcomes of these public tenders in the region.

In Chapter 10, I examine the installation of a Chinese military-connected space research base in Neuquén, Argentina, and highlight the discussion of this project's scientific and technological nature. This case study may be perceived as different from the previous ones because it does not involve direct competition between suppliers. Still, I argue that it is an inescapable case to illustrate perceptions of security and risk in the region. It also connects with the overarching argument of this book because of the reaction of the US and other countries in Latin America. I also explore space research as a form of technology transfer.

With these cases, I intend this book to be an insightful introduction to the intersection of technology, governance, and development. It explores both current and historical cases, making it a resource for future regional studies or for comparing findings across the globe. I also hope to continue to discuss and update the findings of this research as the region evolves and geopolitical conditions change in the future.

Looking Ahead in Technology Transfer and IRs Research

Chapter 11 is devoted to discussion and analysis. The initial conclusions of this book show that when only economic factors are considered and when Chinese firms cooperate with local and more domestic firms, they have a significant advantage in the region. When geopolitical, security, and occasionally value-related issues come into play, firms from other origins gain an advantage over PRC competitors. This conclusion contradicts the narrative and notion that only the best economic offer is always selected. As this book acknowledges, the role of beauty contests as bidding schemes has had its fair share of influence in Latin American politics, and even more so when it comes to telecommunications.

Values and soft power appear as factors in the establishment of national and international norms, both in the case of narratives or arguments advanced by authoritarian countries[83] and in the case of norms shared across the globe[84] (often based on a set of more "liberal" values). After exploring this with sources but also consulting documents published by regional and national public bodies, from the OAS to the telecommunications institutions in each country, it became clear to me that the implementation of national laws in Latin America takes into account international discussions, and that the standards and models cited by sources and documentation often reflect the systems of democratic and developed nations over competing powers such as Russia or China. Even more, some bidding processes directly mention specific foreign frameworks and international standards to be replicated in their processes.

The overall outcome of this book shows the deep linkages between technology and IRs and political discussions. The choice of a provider in the technology sector not only affects citizens from an infrastructure point of view but also impacts their ability to connect and develop in the context of ever-expanding digital media, where these factors affect not only their economic prospects but also their participation in globalization and the coming digital revolution.

Notes

1 Ortega y Gasset (2013), p. 1.
2 Ortega y Gasset (2013), pp. 20–37.
3 Reese (2018).
4 Ferguson (2012).
5 Boehme-Neßler (2011a).
6 Galeano and Belfrage (1997).
7 *World Internet Users Statistics and 2023 World Population Stats* (2023).
8 Drezner (2019); Weiss (2005); Krishna-Hensel (2010); Eriksson and Newlove-Eriksson (2021).
9 Weiss (2005); Hoijtink and Leese (2019).
10 Morgenthau (1948), p. 13.
11 Gilpin (1981).
12 Nye (2011).
13 Nye (2008), pp. 94–96.
14 Weiss (2005).
15 Heim and Miller (2020), p. 7.
16 Nye (2011).
17 McCarthy (2015), p. 20.
18 Skolnikoff (1994), p. 10.
19 Skolnikoff (1994), p. 14.
20 Eriksson and Newlove-Eriksson (2021), pp. 5–14.
21 Eriksson and Newlove-Eriksson (2021), p. 6.
22 Ericsson and Giacomello (2007), pp. 179–181.
23 Ericsson and Giacomello (2007); Eriksson and Newlove-Eriksson (2021).
24 Saliu (2020), pp. 71–72.
25 Coetzee and Hudson (2012), online.
26 Rose (1998), p. 151; Gvalia et al. (2019), pp. 21–22.
27 Ripsman (2009), pp. 170–173.
28 Gourevitch (1978), p. 894.
29 Putnam (1988), p. 433
30 Gerschenkron (1962), pp. 6–10.
31 Gerschenkron (1962), pp. 6–10.
32 Drezner (2019).
33 For example, the Japanese government has made "Society 5.0" a key part of its development strategy to improve the quality of life for its citizens. (*Society 5.0 - 科学技術政策 - 内閣府*, n.d.).
34 J. S., personal communication (June 3, 2024).
35 Tiezzi (2020) and Lee (2020).
36 The use of the term "trade war" is widely contested, but it is a well-known and used concept to describe the process of economic measures and countermeasures between the two countries.

37 Dollar (2022).
38 Other authors that have referenced this issue include Zhou (2019), who builds theory on hegemonic powers and the role of China or Medeiros (2019), who links the economic rivalry to political are more profound consequences regarding the relationship between the two countries.
39 Friedberg (2005).
40 Tekir (2020); Kaska et al. (2019).
41 Throughout 2017 and 2018, numerous European governments-including the "Big Three" recipients of Chinese capital, Germany, France, and the United Kingdom (UK)-proposed or passed new laws that increase scrutiny of foreign mergers and acquisitions for potential national security risks (Hanemann et al., 2019, p. 9).
42 Brian Nougrères (2016), p. 147.
43 Kimbal et al. (2023), online.
44 Chan (2015). Article 3.
45 Comparing this behavior with Rogers's theory, if these countries were users, they would rather be in the latter phase of the process between the early majority (the need) and the late majority (the reluctance) (Rogers, 1962, p. 22).
46 Carrillo (2017).
47 Bojalil et al. (2019).
48 *World Internet Users Statistics and 2023 World Population Stats* (2023).
49 Fu (2021), online.
50 Haerpfer et al. (2022), Q. 163.
51 Comisión Económica para América Latina y el Caribe or ECLAC in English (2017).
52 Shoujun et al. (2018), p. 263.
53 Colombo et al. (2021), pp. 105–107.
54 OECD (2019), pp. 41–42.
55 Venturini and Garay (2021).
56 Kobek and Caldera (2016).
57 Aguilar Antonio (2021).
58 Some case studies are base solely in these testimonies and bibliographical reviews, while other cases incorporate text-based quantitative analysis.
59 D. G. and A. O., personal communication (March 8, 2023); A. G. Z., personal communication (March 13, 2023); R. M. L., personal communication (March 16, 2023).
60 T. P., personal communication (February 27, 2023).
61 P. S., personal communication (January 18, 2023); D. G. and A. O., personal communication (March 8, 2023).
62 J. S., personal communication (June 3, 2024).
63 Dunn Cavelty (2007), pp. 85–86.
64 Eriksson and Giacomello (2007).
65 Oxford Analytica (2019), online.
66 Hurwitz (2014), p. 334.
67 Countries that share political values, economic systems or an overall view or the international order.
68 McKune and Ahmed (2018), p. 3838.
69 Dunn Cavelty (2007), p. 85.
70 Portugal-Perez et al. (2009), p. 18.
71 Zwanenberg et al. (2013), p. 3.
72 Eriksson and Giacomello (2007), pp. 11–16.
73 Rose (1998), pp. 151–153.
74 Boehme-Neßler (2011a), pp. 2–4.
75 Wikander (2000), p. 650.
76 Gilfillan (1965).
77 Saxonhouse (1974).

78 Reese (2018).
79 Schwab (2017).
80 Zakaria (2020), pp. 103–105.
81 Zakaria (2020), pp. 117–120.
82 Hiebert (2022), online.
83 McKune and Ahmed (2018), p. 3838.
84 Hurwitz (2014), pp. 226–330.

Bibliography

Aguilar Antonio, J. M. (2021). Retos y oportunidades en materia de ciberseguridad de América Latina frente al contexto global de ciberamenazas a la seguridad nacional y política exterior. *Estudios Internacionales*, *53*(198), Article 198. https://doi.org/10.5354/0719-3769.2021.57067

Boehme-Neßler, V. (2011a). Caught Between Technophilia and Technophobia: Culture, Technology and the Law. In V. Boehme-Neßler (Ed.), *Pictorial Law: Modern Law and the Power of Pictures* (pp. 1–18). Springer. https://doi.org/10.1007/978-3-642-11889-0_1

Boehme-Neßler, V. (2011b). Cultural Technology and the Law – The Example of Writing. In V. Boehme-Neßler (Ed.), *Pictorial Law: Modern Law and the Power of Pictures* (pp. 19–49). Springer. https://doi.org/10.1007/978-3-642-11889-0_2

Bojalil, P., Egan, M., & Vela-Treviño, C. (2019, February 12). *Data Privacy Reform Gains Momentum in Latin America*. Abierto al Público. https://blogs.iadb.org/conocimiento-abierto/en/data-privacy-reform-gains-momentum-in-latin-america/

Brian Nougrères, A. (2016). Data Protection and Enforcement in Latin America and in Uruguay. In D. Wright & P. De Hert (Eds.), *Enforcing Privacy: Regulatory, Legal and Technological Approaches* (pp. 145–180). Springer International Publishing. https://doi.org/10.1007/978-3-319-25047-2_7

Carrillo, A. M. (2017). Vaccine Production, National Security Anxieties and the Unstable State in Nineteenth- and Twentieth-century Mexico. In C. Holmberg, S. Blume, & P. Greenough (Eds.), *The Politics of Vaccination* (pp. 121–147). Manchester University Press. https://www.jstor.org/stable/j.ctt1wn0s1m.11

Chan, C. L. (2015). Fallen Behind: Science, Technology and Soviet Statism. *Intersect: The Stanford Journal of Science, Technology, and Society*, *8*(3), Article 3. https://ojs.stanford.edu/ojs/index.php/intersect/article/view/691

Coetzee, E., & Hudson, H. (2012). Democratic Peace Theory and the Realist-Liberal Dichotomy: The Promise of Neoclassical Realism? *Politikon*, *39*(2), 257–277. https://doi.org/10.1080/02589346.2012.683942

Colombo, S., López, M. P., & Vera, N. (2021). Tecnologías emergentes, poderes en competencia y regiones en disputa: América latina y el 5G en la contienda tecnológica entre China y Estados Unidos. *Estudos Internacionais: revista de relações internacionais da PUC Minas*, *9*(1), Article 1. https://doi.org/10.5752/P.2317-773X.2021v9.n1.p94

Comisión Económica para América Latina y el Caribe. (n.d.). La Franja y la Ruta es una propuesta civilizatoria de conectividad y prosperidad compartida: CEPAL [Text]. ECLAC; Comisión Económica para América Latina y el Caribe. Retrieved February 8, 2024, from https://www.cepal.org/es/comunicados/la-franja-la-ruta-es-propuesta-civilizatoria-conectividad-prosperidad-compartida-cepal

Dollar, D. (2022). U.S.-China Trade Relations in an Era of Great Power Competition. *China Economic Journal*, *15*(3), 277–289. https://ideas.repec.org//a/taf/rcejxx/v15y2022i3p277-289.html

Drezner, D. W. (2019). Technological Change and International Relations. *International Relations*, *33*(2), 286–303. https://doi.org/10.1177/0047117819834629
Dunn Cavelty, M. (2007). Securing the Digital Age: The Challenges of Complexity for Critical Infrastructure Protection and IR Theory. In *International Relations and Security in the Digital Age* (pp. 85–105). Routledge. https://doi.org/10.4324/9780203964736
Eriksson, J., & Giacomello, G. (Eds.). (2007). *International Relations and Security in the Digital Age*. Routledge. https://doi.org/10.4324/9780203964736
Eriksson, J., & Newlove-Eriksson, L. M. (2021). Chapter 1: Theorizing Technology and International Relations: Prevailing Perspectives and New Horizons. In *Technology and International Relations*. Edward Elgar. https://www.elgaronline.com/edcollchap/edcoll/9781788976060/9781788976060.00007.xml
Ferguson, N. (2012). *Civilization: The West and the Rest*. Penguin Books.
Friedberg, A. L. (2005). The Future of U.S.-China Relations: Is Conflict Inevitable? *International Security*, *30*(2), 7–45. https://doi.org/10.1162/016228805775124589
Fu, Y. (2021). *Attitudes Towards Science, Technology, and Surveillance in 49 Countries · Yiqin Fu*. Github. https://yiqinfu.github.io/posts/global-science-attitudes-2021/
G. D., & O. A. (2023, March 8). *Interview with Authorities at the Telecommunications Agency, Chile*. [In Person].
G. Z., A. (2023, March 13). *Interview with an Inter-American Development Bank Representative* [In Person].
Galeano, E., & Belfrage, C. (1997). *Open Veins of Latin America: Five Centuries of the Pillage of a Continent*. NYU Press. https://www.jstor.org/stable/j.ctt9qfzgp
Gerschenkron, A. (1962). *Economic Backwardness in Historical Perspective*. The Belknap Press of Harvard University Press, WorldCat.
Gilfillan, S. C. (1965). Lead Poisoning and the Fall of Rome. *Journal of Occupational Medicine*, *7*(2), 53–60. https://www.jstor.org/stable/45016005
Gilpin, R. (Ed.). (1981). Growth and Expansion. In *War and Change in World Politics* (pp. 106–155). Cambridge University Press. https://doi.org/10.1017/CBO9780511664267.005
Gourevitch, P. (1978). The Second Image Reversed: The International Sources of Domestic Politics. *International Organization*, *32*(4), 881–912. https://www.jstor.org/stable/2706180
Gvalia, G., Lebanidze, B., & Siroky, D. S. (2019). Neoclassical Realism and Small States: Systemic Constraints and Domestic Filters in Georgia's Foreign Policy. *East European Politics*, *35*(1), 21–51. https://doi.org/10.1080/21599165.2019.1581066
Haerpfer, C., Inglehart, R., Moreno, A., Welzel, C., Kizilova, K., Diez-Medrano, J., Lagos, M., Norris, P., Ponarin, E., & Puranen, B. (2022). World Values Survey Wave 7 (2017–2022) Cross-National Data-Set (Version 4.0.0) [Dataset]. [object Object]. https://doi.org/10.14281/18241.18
Hanemann, T., Huotari, M., & Kratz, A. (2018). *Chinese FDI in Europe: 2018 Trends and Impact of New Screening Policies*. Merics & Rhodium Group (Updated in 2019).
Heim, J., & Miller, B. (2020). *Measuring Power, Power Cycles, and the Risk of Great-Power War in the 21st Century*. RAND Corporation. https://doi.org/10.7249/RR2989
Hiebert, K. (2022). *Democracies Are Dangerously Unprepared for Deepfakes—Centre for International Governance Innovation*. Center for International Governance Innovation. https://www.cigionline.org/articles/democracies-are-dangerously-unprepared-for-deepfakes/
Hoijtink, M., & Leese, M. (Eds.). (2019). *Technology and Agency in International Relations*. Taylor & Francis. https://library.oapen.org/handle/20.500.12657/37352
Hurwitz, R. (2014). The Play of States: Norms and Security in Cyberspace. *American Foreign Policy Interests*, *36*(5), 322–331. https://doi.org/10.1080/10803920.2014.969180

Kaska, K., Beckvard, H., & Minárik, T. (2019). *Huawei, 5G and China as a Security Threat*. Nato Cooperative Cyber Defence Centre of Excellence, CCDCOE.

Kimbal, E., Jo, J. A., Chen Weiss, J., Nye, J., Turpin, M., Kim, P., & Hass, R. (2023). *Should the US Pursue a New Cold War with China?* Brookings. https://www.brookings.edu/articles/should-the-us-pursue-a-new-cold-war-with-china/

Kobek, L. P., & Caldera, E. (2016). Cyber Security and Habeas Data: The Latin American Response to Information Security and Data Protection. *Oasis*, *24*, 109–128. https://www.redalyc.org/journal/531/53163716007/html/

Krishna-Hensel, S. F. (2010). *Technology and International Relations*. Oxford Research Encyclopedia of International Studies. https://doi.org/10.1093/acrefore/9780190846626.013.319

Latin America Seeks Improved Cybersecurity Governance. (2019). Oxford Analytica. https://doi.org/10.1108/OXAN-DB241858

Lee, N. T. (2020). Navigating the U.S.-China 5G Competition. *Global China, Brookings*.

M. L., R. (2023, March 16). *Interview with a Former High-level Politician, Colombia.* [Online].

McCarthy, D. R. (2015). Power and Information Technology: Determinism, Agency, and Constructivism. In D. R. McCarthy (Ed.), *Power, Information Technology, and International Relations Theory* (pp. 19–42). Palgrave Macmillan. https://doi.org/10.1057/9781137306906_2

McKune, S., & Ahmed, S. (2018). Authoritarian Practices in the Digital Age| The Contestation and Shaping of Cyber Norms through China's Internet Sovereignty Agenda. *International Journal of Communication*, *12*, Article 0. https://ijoc.org/index.php/ijoc/article/view/8540

Medeiros, E. S. (2019). The Changing Fundamentals of US-China Relations. *The Washington Quarterly*, *42*(3), 93–119. https://doi.org/10.1080/0163660X.2019.1666355

Morgenthau, H. (1948). *Politics Among Nations* (1st. (Digital)). A. A. Knopf.

Nye, J. S. (2008). Public Diplomacy and Soft Power. *The Annals of the American Academy of Political and Social Science*, *616*(1), 94–109. https://doi.org/10.1177/0002716207311699

Nye, J. S. (2011). *The Future of Power* (1st ed). PublicAffairs; WorldCat.

OECD. (2019). The Road to 5G Networks: Experience to Date and Future Developments. OECD. https://doi.org/10.1787/2f880843-en

Ortega y Gasset, J. (1964). Meditación de la técnica y otros ensayos sobre ciencia y filosofía. In *Obras Completas: Vol. V (Sixth)*. Revista Occidente. https://dialnet.unirioja.es/servlet/libro?codigo=156966

Ortega y Gasset, J. (2013, Original published in 1939). *Meditación de la técnica y otros ensayos sobre ciencia y filosofía*. Ortega y Gasset foundation; Public Domain. https://ortegaygasset.edu/biblioteca/

Portugal-Perez, A., Reyes, J.-D., & Wilson, J. S. (2009). Beyond the Information Technology Agreement: Harmonization of Standards and Trade in Electronics. *Policy Research Working Papers*. https://doi.org/10.1596/1813-9450-4916

Putnam, R. D. (1988). Diplomacy and Domestic Politics: The Logic of Two-Level Games. *International Organization*, *42*(3), 427–460. https://www.jstor.org/stable/2706785

Reese, B. (2018). *The Fourth Age: Smart Robots, Conscious Computers, and the Future of Humanity*. Atria Books.

Ripsman, N. M. (2009). Neoclassical Realism and Domestic Interest Groups. In S. E. Lobell, N. M. Ripsman, & J. W. Taliaferro (Eds.), *Neoclassical Realism, the State, and Foreign Policy* (1st ed., pp. 170–193). Cambridge University Press. https://doi.org/10.1017/CBO9780511811869.006

Rogers, E. M. (1962, ed. 1983). *Diffusion of Innovations* (3rd ed.). Free Press, Collier Macmillan.

Rose, G. (1998). Neoclassical Realism and Theories of Foreign Policy. *World Politics*, *51*(1), 144–172. https://www.jstor.org/stable/25054068

S., J. (2024, June 3). *Telecommunications Consultant with Direct Connections with the International Market* [Online].

S., P. (2023, January 18). *Expert on China's Relations with Latin America* [Online].

Saliu, H. (2020). The Evolution of the Concept of Public Diplomacy from the Perspective of Communication Stakeholders. *Medijska Istraživanja, 26*(1), 69–86. https://doi.org/10.22572/mi.26.1.4

Saxonhouse, G. (1974). A Tale of Japanese Technological Diffusion in the Meiji Period. *The Journal of Economic History, 34*(1), 149–165. https://doi.org/10.1017/S0022050700079687

Schwab, K. (2017). *The Fourth Industrial Revolution, Foreign Language Books*. Portfolio Penguin. https://law.unimelb.edu.au/__data/assets/pdf_file/0005/3385454/Schwab-The_Fourth_Industrial_Revolution_Klaus_S.pdf

Shoujun, C., Zheng, Z., Baiyi, W., Dongzhen, Y., Jingsheng, D., Jianmin, Y., Zhenxing, S., Shixue, J., Yinghua, Z., Fan, Z., Haibin, N., Xiaodai, X., Cunhai, G., & Shuangrong, H. (2018). China Y La Infraestructura En América Latina Desde La Perspectiva De La Diplomacia Económica. In W. Baiyi (Ed.), *Pensamiento social chino sobre América Latina* (pp. 261–290). CLACSO. https://doi.org/10.2307/j.ctvnp0jw3.14

Skolnikoff, E. B. (1994). *The Elusive Transformation: Science, Technology, and the Evolution of International Politics*. Princeton University Press.

Society 5.0—科学技術政策—内閣府. (n.d.). Retrieved February 6, 2024, from https://www8.cao.go.jp/cstp/society5_0/

Tekir, G. (2020). Huawei, 5G Network and Digital Geopolitics. *International Journal of Politic and Security, 2*, 113–135.

Tiezzi, S. (2020). *China's Bid to Write the Global Rules on Data Security*. The Diplomat. https://thediplomat.com/2020/09/chinas-bid-to-write-the-global-rules-on-data-security/

Venturini, J., & Garay, V. (2021). *Reconocimiento facial en América Latina: Tendencias en la implementación de una tecnología perversa*. Al Sur. https://www.alsur.lat/sites/default/files/2021-11/ALSUR_Reconocimiento_facial_en_Latam_ES.pdf

Weiss, C. (2005). Science, Technology and International Relations. *Technology in Society, 27*(3), 295–313. https://doi.org/10.1016/j.techsoc.2005.04.004

Wikander, Ö. (2000). The Roman Empire. In *Handbook of Ancient Water Technology* (pp. 649–660). Brill. https://doi.org/10.1163/9789004473829_031

World Internet Users Statistics and 2023 World Population Stats. (2023). https://www.internetworldstats.com/stats.htm

https://ourworldindata.org/grapher/share-of-individuals-using-the-internet

Zakaria, F. (2020). *Ten Lessons for a Post-Pandemic World*. W W Norton. https://wwnorton.com/books/9780393542134

Zhou, J. (2019). Power Transition and Paradigm Shift in Diplomacy: Why China and the US March towards Strategic Competition? *The Chinese Journal of International Politics, 12*(1), 1–34. https://doi.org/10.1093/cjip/poy019

Zwanenberg, P. van, Ely, A., & Smith, A. (2013). *Regulating Technology: International Harmonization and Local Realities*. Routledge.

2 Technology's Cultural Influence on Global Dynamics

From History to Modern Debates on Innovation

Technology transfer and the geopolitical consequences of states' decisions in Latin America amid a growing rivalry between the US and the PRC are critical issues. Nevertheless they can't be understood without examining the importance of technology for cultural and economic development. This book is based on the explicit and overall shared assumption that technology and science are crucial to humanity's development[1] and have been shown to impact culture, society, and historical outcomes profoundly. It is only natural to connect the potential of weapons and warfare with how a group demonstrates power. However, technologies also have an impact on individuals at a developmental level. For instance, if a society has access to scientific, infrastructure, or medical advancements, it gains an advantage over others, which can result in significant differences in output. As a result, the impact of technology includes practical considerations such as living standards and broader societal issues. It also involves interpreting problems and solutions through various ideological models and achieving different developmental outcomes based on a group's access to innovation.

This chapter explores the consequences of technological development by examining technology's historical and philosophical significance alongside crucial factors contributing to the comparative advantage of certain nations over others. Exploring this relationship can also help readers understand why states place such importance on technology transfer and diffusion.

By examining the Latin American context, I also propose that it is possible to frame the perceived lack of international importance of technology creation within the region and the incentives that lead states in this region to consider incorporating innovations produced abroad and to negotiate technology transfer, even in cases where the management and use of sensitive data is involved. This can also be analyzed in the following chapters when it comes to decision-making and the pressures on governments to manage both domestic needs and international norms and geopolitical phenomena.

Finally, this chapter elucidates the relevance of technology to the analysis of power within international relations research and its implications for the contemporary global landscape. It also features historical examples and comparative studies between different periods and socio-political systems where technology was, and continues to be, developed.

DOI: 10.4324/9781003489450-3

The Circular Relationship between Technology and Development

The pivotal role of technology in shaping humanity's cultural and civilizational outputs is indisputable. From the Greek *τέχνη* or *techne*, meaning "art, skill, craft," and logos, meaning "reasoning and knowledge,"[2] technology encompasses more than just machines and software; it also refers to how humans use their accumulated wisdom to shape their environments. Knowing how to light a fire or using tools to work the land or as weapons marked a difference between *Homo Sapiens* and its relatives.[3] Strength and nature could be overcome by utilizing knowledge and ingenuity through inventions and relative command of natural forces.

Technology also defines historical periods. Agriculture—considered by some as a technology-related feat due to the knowledge and tools used to master it—marked the beginning of sedentary life and, therefore, settlements and civilization. Under that same logic, the development of language, writing, and the press would be considered the stepping stones in the history of modern inventions. As a parenthesis and to narrow down a definition for this book, I assume there is a need to separate traditional notions of technology from what is currently considered or related to innovation. The internet, as well as the debates on current affairs such as the use of artificial intelligence, would not exist without the creation of means of communication and languages that can be learned by humans and machines.[4] Nevertheless, I do not consider that the trade of paper, wood, or seeds—technology and knowledge products—can be regarded as technology transfer today.

In any case, technological advancement has emerged as a significant factor determining competition among various groups, from tribes to modern states. Ortega y Gasset suggests that technology is how humans adapt to their environment and is closely related to development. It is also how a society simplifies processes to meet its needs.[5] In that regard, "technical history," or the societal and cultural factors that determine certain dispositions toward innovation, can be directly connected to the influence of a state, culture, or society. It is more of a two-way process since certain social, political, and economic conditions are required to develop technology,[6] and consequently, technology provides a comparative advantage.

Several factors influence this cycle or upward spiral. First, the economy of a nation, empire, or any social group determines how many resources they can invest in development. If decision-makers choose the right path, innovation will give the said group an advantage over others.[7] Philosophy is, for others, a critical element that separates innovators and affects both culture and statehood's relationship with technology.[8] Thus, while laws may provide the structure for development, the innovative capacity of a society often determines the nature and direction of its legal framework regarding technology. To create said laws, policymakers must interpret philosophical proposals in the form of ideologies, and there needs to be a cultural and social change that incites those processes. Hierarchies, power structures, governance systems, and even social relations are, in many ways, intertwined with technological capacity.[9]

Nevertheless, limiting this process only to a circular or bidirectional societal developmental environment is not enough. Under that logic, only a handful of groups would historically develop over others without any form of hierarchical change or possibility for competition. Societies—as empires—can rise and fall, and institutions forged out of philosophical principles can be shared and implemented across the globe. As historian Niall Ferguson elaborates, a series of institutional frameworks developed worldwide and based on ideas that are sometimes closely related to the Enlightenment were vital in determining how science—and medicine in a relevant context—developed as an art and tool since the late 16th century. Societies in and out of the "West"[10] have subsequently been able to apply and adapt those institutions to their legal and societal frameworks.[11] As a result, innovation due to state- or group-level policies is mainly related to a handful of nations, but successful outliers emerge every decade around the world.

The opposite can also be true. Societies that at one point enjoyed a prevalent position in terms of development and culture can lose said "advantage" over time when the conditions for innovation change or technologies fail to answer specific needs. In addition, scientific and technological improvements do not always lead to industrialization, social benefits, or international relative superiority since not all technologies were traditionally relevant regarding power leverage.

Perhaps the most mentioned example to convey this non-static international power, as well as the formation and change of culture, is the situation of China around the 14th and 15th centuries.[12] The country was arguably among the most advanced in the world. Still, that progress did not lead to enlightenment and industrialization,[13] allowing European countries to dominate in innovation and the philosophy surrounding them. The reasons for China's technological decline are often connected to bureaucratic, political, and economic changes that undermined innovation and development and, therefore, affected its capacity for technological progress.[14] In other words, during the Ming Dynasty, China peaked in comparative technological innovation, and gunpowder became a central technological achievement internationally. By the 16th century, the state had started to implement more absolutist policies, and it declined into what is sometimes called the beginning of a "regression" process—which reached its lowest point in later centuries and with colonial powers expanding to the East.[15] In that period, a scholarship of traditionalist national ideologists was emphatic in dismissing the introduction of Western inventions, arguing that most of these had originated in China. However, some managed to land through contact with, for example, the Jesuits.[16] That situation changed again, but not sooner than the late 20th century, when policy, economic, and global conditions enabled mainland China's current and scientific improvements. This has given the PRC advantages in, for example, space competition at an international level.[17] In other areas, such as transportation, telecommunications, or medicine, the country has also been highlighted as a competitor to the US and other technologically advanced nations.[18] In the following chapters, some of these examples are better explained, as well as the role the Chinese Communist Party plays in the dissemination of specific technologies of interest.

On the other side of the strait, Taiwan—with a considerably different political system—has also implemented policies and national strategies that, in coordination with the private sector and academia, made the island a semiconductor manufacturer now at the epicenter of geopolitical debate.[19] After the Second World War, the island was de-colonized by the Japanese, but the Chinese civil war was still an open conflict on the mainland. Agriculture had long been Taiwan's main economic activity, and few could argue that there was a pushing technology-friendly culture brewing across society.[20]

In 1971, the UN published its Resolution 2758, which, among other matters, proposed the "Restoration of the lawful rights of the People's Republic of China in the United Nations" as the only legitimate state to represent China's interests and expelled Chiang Kai-shek from said position.[21] Two years after it lost its international recognition, the Taiwanese government established the Industry Technology Research Institute (ITRI) and the Development Fund to promote the growth and development of the semiconductor industry decisively. By the late 1990s, Taiwan was the fourth largest silicon chip producer behind Korea and Japan and the first, the US.[22] Today, the debate over a potential invasion of Taiwan cannot be divorced from its technological and strategic significance in global supply chains and innovation.[23] The Taiwanese experience leads to two conclusions. Firstly, there is a solid but debatable relationship between the UN resolution and Taiwan's policy decision. Secondly, the nation's geopolitical relevance is interconnected to its technological culture and history. These examples demonstrate that innovation cannot be viewed solely as a means to find solutions to everyday problems.

Another well-known example of cultural and political changes that hindered or promoted development throughout the centuries is Japan's two-century isolationism. During the 16th century, firearms—introduced in 1543 to the country—were instrumental in Oda Nobunaga's plans for Japanese unification, later achieved by his successor Toyotomi Hideyoshi,[24] which extended to the Tokugawa shogunate. A few years later, firearms were also instrumental in the Korean invasion (1592–1598) and against Chinese troops of up to 200,000 soldiers.[25] But this quickly changed. By 1607, economic, cultural, and political reasons had led the country back to traditional weaponry,[26] and in 1640, the *Sakoku*, or the process of Japanese isolation—which also brought overall peace[27]—left society relatively stagnated, when compared to the industrialization processes that were being undertaken in the west.[28]

Western science was heavily banned from the country in the 18th century. Nevertheless, science did enter the country, first due to translations of Dutch texts—some commissioned—through traded books, and increasingly by Japanese who criticized traditional Chinese scholarship compared to the infiltrated Western Science.[29] In other words, a cultural context of curiosity and critical thinking established the basis for letting modern thought into the country. The Meiji Restoration in 1868 marked the realization of the comparative advantages of technology and industrialization in Western nations, which had been two centuries in the making while Japan was under the *Sakoku*.[30] The country learned to adapt quickly, and that process was supported by a program of modernization that included scientific, political, and economic reforms.[31] The results of this policy, as well as the philosophy and the changes in living standards, can still be seen to this date.

I must emphasize that many societies and states have utilized technological advantages to address perceived weaknesses in areas such as economics, military, science, medicine, and academia. The shift from an economy based on natural resources to one based on innovation is often attributed to economic strategy, strategic policy, and a robust scientific culture and community. However, only a few analyses I came across during my exploratory research for this book consider individual incentives that might influence how citizens, bureaucrats, and policy-makers change their vision toward how technology should be handled.[32]

Finally, the philosophical framework in which technology develops is also relevant as it shapes policies and laws that rule over the control or creation of technology.[33] The idea of humanity controlling its surroundings and its power over nature is shared in this rhetoric, as in some religious views on the exceptionalism of humankind.[34] In contrast, notions on the ethics of technological development are the center of much debate. The main reason for this is whether technology is produced with a specific objective that has a moral weight to it[35] or if technology is relatively neutral. The user–in this case, a country's leadership–is evaluated in terms of intentions and outcomes. In some examples, the answer is more evident than in others. Nuclear power is capable of unmeasurable destruction, but it also powers cities with a lower carbon emission footprint than other sources. Artificial intelligence (AI) is both feared and seen as an indispensable asset for the future. Even apparently "simpler" inventions, such as the white phosphorus used to top matches, have seen domestic and military use.[36]

In this sense, philosophy and thought are funneled into ideologies that instrumentalize those ideas, which later impact the development and diffusion of innovation. When classical liberalism became politically relevant, as well as the emergence of the scientific method and drastic changes in work culture following the Enlightenment, the scale and volume of innovation skyrocketed. Other movements, such as Marxism, emerged as a critique of the outcomes of the political and societal changes in the continent, although only sometimes opposed to innovation itself. These ideas proposed different interpretations and projections of how technologies affect humans and the debates over how fast and why are new inventions created continue to this date.[37]

Technology Philosophy and Development in Latin America

Some historians examining the process of Latin American colonization emphasize that the region's economic conditions and the way colonization unfolded are vital reasons Latin America has different technological development compared to its neighbors in North America.[38] Eduardo Galeano is often cited for his view on regional exploitation and imperialism. In his most influential work, "Open Veins of Latin America: Five Centuries of the Pillage of a Continent," the author portrays how elites and industries have been more concerned with extracting natural resources—deriving in a society where the economy is foremost based on "what can be sold" to industrial powers,[39] in the form of mercantilism. Latin American scholars have long relied on and replicated this narrative to justify the lack of regional industrialization and development. Still, alternative theories that explore development through institutional frameworks indicate that there are internal factors—such as property distribution and religion—that also contributed to the region's particular lack of scientific and technological curiosity that led to the differences between the US and the southern parts of the continent.[40]

It is necessary to add that these interpretations often overlook the existence of outliers such as Argentina. The country had an outstanding development during the 19th century, and due to institutional changes—based on a set of liberal ideas coherent with the Enlightenment and anti-absolutism,[41] it became one of the world's largest and most prosperous economies. In 1913, Argentina's GDP per capita stood at US$470, France's was US$400, and Japan's stood at US$90.[42] The country had all the means to become a global superpower. Still, due to institutional changes, it declined consistently since the early 1900s, and in 2023, its economy was US$13,650, or six times smaller than that of the US.[43] The reasons for the country's decline are often related to the ideological changes of the 1900s and populism.

In the 20th century, researchers focused on the contradictory situation of Latin American states. These countries were rich in resources and had well-educated elites, many of whom studied in Europe. However, local culture and development did not prioritize competition in technological improvements; instead, they resorted to incorporating them from abroad, with counted exceptions. Some schools of thought have elaborated on the relationship between Latin America's economic development and the causes of its complex adaptation to technology. Still, the number of existing "technology historians" and theorists in the region remains limited to a few,[44] as does the extended academia on international relations and technological production, imports, and adaptations.

During the 1950s, the idea of "dependency" was extensively used to express the problem of the region's lack of industrialization,[45] as well as the lack of endogenous innovation and scientific research. That line of reasoning was born within the United Nations Economic Commission on Latin America (ECLAC), such as Raúl Prebisch and Osvaldo Sunkel, and greatly influenced academics looking for solutions to the region's shared problems in the matter.[46] Their school of thought has also influenced the interpretation of technology transfers to the Latin American region. Investors and technology owners are perceived as dominant, together with states that strive to attract capital and foreign innovation.[47]

Despite these countries being early adapters of foreign technologies—and despite the direct relation between investment in R&D and economic prosperity,[48] the region has not achieved the same success as other "recent" technologically advanced nations. However, I agree with the commonly shared argument that not all economically prosperous countries are necessarily innovators in the technological sense. Nations can become rich by industrializing their processes to refine and utilize natural resources, and this has been the idea behind some of the policies across some Latin American nations. Also, governments in the South Cone have historically preferred financial and trade-related activities. Nevertheless, institutions and culture affect nations' outputs, and the reliance on natural resources and concentrated exports is potentially threatening.

Chile's central wealth during the 19th century is closely related to such trade policies[49] and the establishment of a solid institutionality that looked outwards.[50] Understandably, Chile's ports saw a decline in traffic upon the opening of the

Panama Canal.[51] Today, the nation relies heavily on copper and lithium exports, and other industries have grown around those activities. The country has, therefore, been able to adapt to scenarios specific to the domestic and international conditions around it. Still, it has also suffered the economic impact of commodity price changes, including saltpeter in the 20th century and copper supercycle changes in recent history. Interestingly, Chile is also one of the primary canal users, which has reduced shipping traffic through its ports.[52]

Venezuela, once among the most stable and democratic countries in the region, has always relied on natural resources for its development—although it behaved in a slightly different way compared to other Latin American countries. The dependency on oil exports was—for an extended period—pointed out as a threat to the country's stability,[53] but that didn't change the government's policies or the behavior of the private sector in a significant way.

The influence of Latin American technology abroad is limited but not unheard of. Countries in the region have a prepared population capable of adapting foreign technology to local environments.[54] Technology companies often work as services, and in areas such as some local actors like Claro México, Entel Chile, Mercadolibre Argentina, or Cornershop Chile have managed to expand to other countries in the same region. These cases will be explored in more detail in their proper chapters. In addition, foreign companies such as Intel in Costa Rica have installed operations as exporters. High levels of education among crucial parts of the population have made it possible for some of these countries to export "independent" services from individuals who work from their local environments to overseas companies or clients. I argue that this creates a grey area on technology transfers due to the thin line between the origin of innovation and its conceptualization.

Technology and Different Notions of Power

The previous review introduced the broad theory of the relationship between innovation and philosophy. It is widely accepted that both elements are crucial for the differentiation and development of societies. Culture is amplified and limited to the available technology; a country's ability to produce scientific and technological outputs is a critical factor in international relations.

Understanding the interaction between culture and technology across various historical periods is instrumental in elucidating the motivations driving countries to intensify their endeavors in technology and innovation. Warfare-related innovation serves as a conspicuous illustration of this phenomenon, wherein technological superiority can significantly influence political outcomes, albeit not invariably guaranteeing absolute victory.[55]

Technologies do not evolve in isolation; knowledge sharing is equally significant to innovation. In certain instances, a society may initiate a discovery or invention, only to be later adapted or repurposed by another group. Conversely, existing scientific knowledge is often leveraged to develop novel methodologies for warfare. Occasionally, a discovery initially intended for military application can be repurposed to enhance other facets of societal development.

The discovery of gunpowder—a widely recognized Chinese invention—is one of these examples. The date of emergence of this combination of chemicals is not known, with some estimations going as far back as 85 A.D., and with publications of a formula dating its beginnings to 1040 A.D.[56] Firearms used that technology as they were developed and improved across empires and kingdoms of Europe, the Middle East, and even India between the 13th and 15th centuries.[57] Artillery was also developed in Western Europe around the 14th century.[58] The acquisition of firearms remained a determinant factor in later conflicts. During the Spanish "Conquista de América" or conquest of the Americas, traditional swords and firearms became a decisive advantage for the outnumbered Spanish *conquistadores*,[59] although diplomacy as well as the transmission of disease from the old world are—to some analysts—as crucial as gunpowder.[60]

During the 20th century, several "arms races" marked both the two World Wars and the subsequent Cold War, when the introduction of the atomic bomb and the idea of mutually assured destruction also changed the way states consider the possible outcomes of a conflict. Due to technological competition and rapid improvement of capabilities that directly impact international relations, arms races have been analyzed under several schemes under the framework of game theories, including the "prisoner's dilemma," "The Richardson model," and other economic models.[61]

When comprehending innovation within the IRs framework, it is crucial to acknowledge the correlation between power and technology[62]. Power is itself a definition that causes debate in the discussion of international relations. Morgenthau, for example, distinguishes political power from physical force[63] while defining the elements of national power in terms of material and non-material resources. Material resources refer to tangible matters like military capabilities and geography, while non-material focus on diplomacy, bureaucracy, or morale.[64] Undoubtedly, both are deeply related to the philosophical and historical variables that derive from technology and innovation.

However, power is commonly understood as the ability to force or convince another entity to take action. It can take different forms, such as economic, military, and political influence, and is directly affected by the development of new technologies or advancements in existing ones.[65] Therefore, advanced technology is consistently perceived as a pivotal determinant of national power and in the broader global order.

A state will strive to acquire technology to enhance its standing and relative power, either to discourage others or to exert influence over an outside party. This is closely tied to international relations and serves as the primary framework for examining many scenarios. For instance, the concept of mutually assured destruction (MAD) must adjust to innovations in nuclear weapons or shifts in policies undertaken by the current actors.[66] Some recent actions taken by China and the US also fall under the notion of technology as a direct military asset. Beijing has extended its production capabilities and has been working on transferring innovation to its military branches, enlarging its global influence. On the other hand, Washington has worked to counter this trend, as it seemingly challenges its dominance.[67]

Nevertheless, when confronted with other forms of power or advantages related to technological dominance, the analysis is not as simple as some make it out to be. Since the end of the Cold War and the introduction of the internet, there has been an increasing underlying complexity of technology as a means for extending international influence—and for explaining multilateral interactions that not only implicate states but also private and "pragmatic" actors.[68] As authors Eriksson and Giacomello highlight:

> Governments are challenged to operate in unfamiliar ways, sharing influence with experts in the IT community, with businesses and with non-profit organizations, because the ownership, operation, and supply of the critical systems are in the hands of a largely private industry, which is diverse, intermixed and relatively unregulated.[69]

This analysis reveals two essential elements. On the one hand, in countries where institutions enjoy freedom and independence and operations are multilateral, governments must leverage their influence in the private sector to turn new technologies into assets in their power structure. On the other hand, it emphasizes the current power competition between digital enterprises and information technology. Since the involved actors, private and public, coordinate based on individual interests, the analysis reflects an anarchic international sphere with a much more intricate actor interaction than traditional conflicts.

I also counterargue that not all governments have the limitation of needing to compromise with the private sector. This gives certain states the advantage of weaponizing technology for military use, surveillance, or soft power[70] strategies. As a result, public diplomacy—a concept that has been in use for more than a century[71] but was coined by Edmund Gullion in 1965—is now central in international relations, especially to define a country's strategies that differ from traditional propaganda but that go beyond conventional diplomacy.[72]

As for weaponizing technology, a clear example of this is the PRC's approach to technological innovation. China's cybersecurity law, effective since 2017, requires companies to provide information to the government with few barriers to secure an individual's data when confronted with what the CCP considers national security.[73] This is complemented by the increasing participation of the Chinese Communist Party in private companies' decisions 73—which implies that the Government has gained effective control of the technology sector, its objectives, and development, at least in the most relevant or sensitive areas. This is part of a broader strategy for company management control. In 2020, All-China Federation of Industry and Commerce (ACFIC) Vice Chairman Ye Qing said in a statement that "we must effectively strengthen party building work in private enterprises and enhance enterprise productivity and competitiveness."[74] These factors are considered when a country decides to limit Chinese companies' participation in its technological infrastructure. In the case of Australia, issues ranging from 5G networks to CCTV cameras have raised concern because of the influence Beijing has over the incumbent companies.[75]

Despite the growing concerns about cybersecurity, discussions about Chinese technologies in Latin America tend to remain neutral. However, Beijing has intensified its efforts to enter this market as part of a public diplomacy strategy. This is accomplished by establishing local research centers, offering scholarships through companies like Huawei, and engaging with local governments. For example, the Chilean government recently collaborated with Huawei to send ten students to learn about digitalization and artificial intelligence—the result of a public call for applications that garnered 1,700 candidates.[76]

Other states also utilize technology as a critical aspect of their international strategy. The dominance of Taiwan in the semiconductor sector is a clear deterrent not only to China but to possible allies that need those supplies for their development.[77] Taiwan's government was closely involved in the institutional process that led to the island's current situation.[78] Although chips are not directly part of Taipei's military power, they provide the island with economic power and both a reputation—for soft power—and a threat for those who only measure a possible conflict in terms of measurable impact.

While the 20th century was dominated by arms races that sometimes derived into consumer technology, the 21st century is characterized by digitalization and the loss of practical borders for businesses, technology production, and communication. In some cases, technology transfer affects innovation capacity on both ends, boosts the economy, and creates further wealth while creating better technological outputs.

When a state can use technology to expand its presence in a foreign territory, it increases the recognition of said country's innovation and, therefore, affects reputational factors. The transfer of particular technology can be accompanied by a series of standards that—to some extent—force the receiving party to comply and continue to acquire technology from similar offerors. In this scenario, a country's power is measured by its ability to coerce and expand its influence by establishing these dependent relationships..

At the same time, two or more governments might decide to use similar systems or to develop specific standards on a homologous path, which can be analyzed as a shared strategy. The European Union's General Data Protection Regulation (GDPR) is one recent example of this. The regulation provided an innovative framework for using personal information across the bloc's jurisdiction. Conversely, countries worldwide that shared similar values or looked to homologize their framework to Europe for economic reasons—began to use that law as the standard for their system. In Latin America, this process was almost instantaneous.[79] Although that regulation is not the same as a product, invention, or specific technology, it does affect how the former are conceptualized if they want to be present in those markets or countries.

With all that has been said, there is an ongoing debate about the extent to which power is measured when a group, state, or company has control over a particular innovation. Traditional schools of thought have addressed these questions, but some gaps still remain. Later in this book, I summarize traditional interpretations from constructivist, liberal, and realist perspectives, and I propose a framework to better utilize existing derivative theories to encapsulate the particularities of the

Latin American scenario and the extent to which middle powers can compete with major powers such as China in the current context.

In this chapter, I provided an overview of the historical conditions under which technology affects human development and give examples of how states use these tools to gain international advantages. This is important for a broad and interdisciplinary understanding of technology transfers across borders and the strategic and philosophical implications of this phenomenon. For Latin American states seeking to gain an advantage by incorporating technologies and modernizing their systems, understanding the relationship between institutions, modernization, culture, and philosophy is crucial. In the following chapters of this book, I explore these interconnections from an international relations perspective, drawing from different deontological frameworks.

Notes

1 Although many separate technology developments from scientific knowledge, I write under the assumption—and support—that they are intertwined when it comes to a social groups' development.
2 Parry (2024), online.
3 By Homo Sapiens I do refer not only to modern humans but also to other close species such as Neanderthals and Denisovans who have also been found to have used tools and control fire to an extent.
4 Reese recognizes four eras that start with the Greek myth of Prometheus and the discovery [and control] of fire and how this allowed Hunter-Gatherers to thrive in inhospitable environments. The second technology he highlights is agriculture, which marked the path for organized settlements and more complex interactions.
Reese (2018), pp. 9–21.
5 Ortega y Gasset (1964), first published in 1939, pp. 332–334.
6 Boehme-Neßler (2011a), pp. 2–4.
7 Boehme-Neßler (2011a), p. 3.
8 Rammert (2007), pp. 6–7.
9 Ortega y Gasset (1964), first published in 1939.
10 The debate of what is "the West" is also relevant. Some accounts consider Latin America as an assortment of developing Western nations—something the countries of the region as most of the time supporters of. Nevertheless, the existence of indigenous movements as well as the differences in outputs between North America and South America have driven some research to claim that the region is geographically western but not ideologically so (Espinosa, 2017, online).
11 Ferguson (2012), pp. 54–57.
12 Ferguson (2012), pp. 90–91.
13 Boehme-Neßler (2011a), p. 2.
14 Ferguson (2012), p. 351.
15 Ferguson (2012), p. 49.
16 Wong (1963), pp. 29–30, 43.
17 Handberg and Li (2006), p. 10.
18 In this book, the concept of "nation" is loosely used as a synonym for country for stylistic purposes. This does not mean that the difference between the two is not acknowledged, but that is not relevant for the purpose of this piece of work.
19 Yuan et al. (1998).
20 Fei (1991), pp. 63–64.
21 UN General Assembly (1972).

22 Ouyang (2006).
23 *The Chips That Make Taiwan the Center of the World* (2022).
24 Brown (1948), p. 238.
25 Brown (1948), p. 240.
26 Boehme-Neßler (2011a).
27 Although it is not the purpose of this book, I believe is important to recognize there is still debate over how isolated Japan really was during these 200 years, considering the opening of "*dejima*" or 出島, as well as its wider presence in Nagasaki (Kazui and Videen, 1982). Debates also examine that the Edo period brought cultural "democratization" in terms of education for some members of lower classes and that literacy increased during this time.
28 Ferguson (2012), p. 92.
29 Nakayama (2009b), 337.
30 Ohno (2019), p. 2.
31 Kim (2007).
32 Ouyang (2006).
33 Boehme-Neßler (2011b).
34 Boehme-Neßler (2011a), p. 3.
35 Carr (2016), pp. 17–19.
36 Reyhani (2007), p. 8.
37 Boehme-Neßler (2011: 1), p. 3.
38 Ferguson (2012), pp. 169–170.
39 Galeano and Belfrage (1997), pp. 2–8, 134.
40 Ferguson (2012), pp. 153–160.
41 Hamilton (2005), p. 528.
42 Hamilton (2005), p. 529.
43 *World Bank Open Data* (online).
44 Osorio (2022).
45 Prebisch (1950).
46 Osorio (2022).
47 Wilner (1981), p. 270.
48 Aali Bujari et al. (2016), pp. 88–89.
49 Dingemans (2022).
50 Varas (1999), p. 11.
51 Santana (2017), online.
52 Forbes (2020), online.
53 Morillo Moreno (2007).
54 Teitel (1981).
55 Rome also enjoyed achievements in their conquests that were facilitated by their technical capabilities, but this did not prevent the downfall of the empire due to a multitude of factors, including Germanic invasions from groups that did not have access to the same innovation levels. During the Vietnam War, it was not technological capabilities but strategic and environmental factors that prevented the United States from imposing its victory in the conflict. The same can be said for some conflicts in the Middle East.
56 Ling (1947), pp. 161–162.
57 Andrade (2016), Chapter 5, pp. 75–77.
58 Andrade (2016), Chapter 6, p. 88.
59 Lynch (2001).
60 Diomedi (2003).
61 Perlo-Freeman (2023).
62 Hoijtink, M., and Leese, M. (Eds.). (2019).
63 Morgenthau (1948), p. 13.
64 Morgenthau (1955), p. 129, as seen in Berenskoetter and Williams (2007), p. 49.

65 Lewis (2022), online.
66 Cimbala (1985).
67 Afshan (2021), p. 93.
68 Hurwitz (2014), pp. 332–334.
69 Eriksson and Giacomello (2007), pp. 179–181.
70 As defined in the introduction of this book, soft power refers to a form of influence that does not rely on military or economic threats but on cultural and reputational factors, and is built up with strategies such as public diplomacy.
71 Some attribute the concept to Woodrow Wilson (Lewis, 2022).
72 Saliu (2020), pp. 71–72.
73 Creemers et al. (2017).
74 Translation from the original Chinese. (*Ye Qing: Promote the Organic Integration of the Party's Leadership System and the Private Enterprise Governance System-All-China Federation of Industry and Commerce*, 2020).
75 Bernot and Smith (2023).
76 Government of Chile (2024).
77 Miller (2022).
78 Ouyang (2006).
79 Carrillo and Jackson (2022), online.

Bibliography

Aali Bujari, A., Venegas Martínez, F., Aali Bujari, A., & Venegas Martínez, F. (2016). Technological Innovation and Economic Growth in Latin America. *Revista Mexicana de Economía y Finanzas*, *11*(2), 77–89. https://www.scielo.org.mx/scielo.php?script=sci_abstract&pid=S1665-53462016000200077&lng=es&nrm=iso&tlng=en

Afshan, S. (2021). Balance of Power in the Era of Technological Globalization. *Pakistan Horizon*, *74*(2–3), 81–101.

Andrade, T. (2016). *The Gunpowder Age: China, Military Innovation, and the Rise of the West in World History*. Princeton University Press. https://doi.org/10.2307/j.ctvc77j74

Berenskoetter, F., & Williams, M. J. (2007). *Power in World Politics*. Routledge.

Bernot, A., & Smith, M. (2023). *Regulating Chinese-Made CCTV Cameras in Australia*. Australian Institute of International Affairs. https://www.internationalaffairs.org.au/australianoutlook/regulating-chinese-made-cctv-cameras-in-australia/

Boehme-Neßler, V. (2011a). Caught between Technophilia and Technophobia: Culture, Technology and the Law. In V. Boehme-Neßler (Ed.), *Pictorial Law: Modern Law and the Power of Pictures* (pp. 1–18). Springer. https://doi.org/10.1007/978-3-642-11889-0_1

Boehme-Neßler, V. (2011b). Cultural Technology and the Law – The Example of Writing. In V. Boehme-Neßler (Ed.), *Pictorial Law: Modern Law and the Power of Pictures* (pp. 19–49). Springer. https://doi.org/10.1007/978-3-642-11889-0_2

Brown, D. M. (1948). The Impact of Firearms on Japanese Warfare, 1543–98. *The Far Eastern Quarterly*, *7*(3), 236–253. https://doi.org/10.2307/2048846

Carr, M. (2016). International Relations Meets Technology Theory. In M. Carr (Ed.), *US Power and the Internet in International Relations: The Irony of the Information Age* (pp. 16–44). Palgrave Macmillan UK. https://doi.org/10.1057/9781137550248_2

Carrillo, A., & Jackson, M. (2022). Follow the Leader? A Comparative Law Study of the EU's General Data Protection Regulation's Impact in Latin America. *ICL Journal*, *15*(2), 177–162. https://www.degruyter.com/document/doi/10.1515/icl-2021-0037/html?lang=en

Cimbala, S. J. (1985). Forever MAD: Essence and Attributes. *Armed Forces & Society*, *12*(1), 95–107. https://www.jstor.org/stable/45304827

Creemers, R., Webster, G., & Triolo, P. (2017). *Translation: Cybersecurity Law of the People's Republic of China (Effective June 1, 2017)*. DigiChina. https://digichina.stanford.edu/work/translation-cybersecurity-law-of-the-peoples-republic-of-china-effective-june-1-2017/

Dingemans, A. F. (2022). Los designios de la política comercial de Chile: Adecuaciones mediante y pragmatismo en las medidas legislativas, 1850–1914. *América Latina en la historia económica, 29*(3). https://doi.org/10.18232/20073496.1314

Diomedi, P. A. (2003). La guerra biológica en la conquista del nuevo mundo: Una revisión histórica y sistemática de la literatura. *Revista Chilena de Infectología, 20*(1), 19–25. https://doi.org/10.4067/S0716-10182003000100003

Eriksson, J., & Giacomello, G. (Eds.). (2007). *International Relations and Security in the Digital Age*. Routledge. https://doi.org/10.4324/9780203964736

Espinosa, E. L. de. (2017). *Is Latin America Part of the West?* Elcano Royal Institute. https://www.realinstitutoelcano.org/en/work-document/is-latin-america-part-of-the-west/

Fei, J. C. H. (1991). Taiwan's Economic Development and Its Relationship to the International Environment. *Asian Affairs, 18*(2), 63–79. https://www.jstor.org/stable/30172335

Ferguson, N. (2012). *Civilization: The West and the Rest*. Penguin Books.

Forbes. (2020, July 24). *Histórico apoyo de Chile con el canal de Panamá*. Forbes Centroamérica. https://forbescentroamerica.com/2020/07/24/historico-apoyo-de-chile-con-el-canal-de-panama/

Galeano, E., & Belfrage, C. (1997). *Open Veins of Latin America: Five Centuries of the Pillage of a Continent*. NYU Press. https://www.jstor.org/stable/j.ctt9qfzgp

Government of Chile. (2024). *Diez estudiantes viajarán a China a potenciar sus habilidades digitales gracias a las Becas Mineduc-Huawei*. Ministerio Secretaría General de Gobierno. https://msgg.gob.cl/wp/2024/09/03/diez-estudiantes-viajaran-a-china-a-potenciar-sus-habilidades-digitales-gracias-a-las-becas-mineduc-huawei/

Hamilton, J. I. G. (2005). Historical Reflections on the Splendor and Decline of Argentina. *Cato Journal, 25*(3), 521–540. https://heinonline.org/HOL/P?h=hein.journals/catoj25&i=536

Handberg, R., & Li, Z. (2006). *Chinese Space Policy: A Study in Domestic and International Politics*. Routledge.

Hoijtink, M., & Leese, M. (Eds.). (2019). *Technology and Agency in International Relations*. Taylor & Francis. https://library.oapen.org/handle/20.500.12657/37352

Hurwitz, R. (2014). The Play of States: Norms and Security in Cyberspace. *American Foreign Policy Interests, 36*(5), 322–331. https://doi.org/10.1080/10803920.2014.969180

Kazui, T., & Videen, S. D. (1982). Foreign Relations during the Edo Period: Sakoku Reexamined. *Journal of Japanese Studies, 8*(2), 283–306. https://doi.org/10.2307/132341

Kim, D.-W. (2007). On Building a Modern Japan: Science, Technology, and Medicine in the Meiji Era and Beyond. *East Asian Science, Technology and Society: An International Journal, 1*(2), 255–258. https://doi.org/10.1215/s12280-007-9016-3

Lewis, J. A. (2022). *Technology and the Shifting Balance of Power*. CSIS. https://www.csis.org/analysis/technology-and-shifting-balance-power

Ling, W. (1947). On the Invention and Use of Gunpowder and Firearms in China. *Isis, 37*(3/4), 160–178. https://www.jstor.org/stable/225569

Lynch, J. (2001). Arms and Men in the Spanish Conquest of America. In *Latin America between Colony and Nation: Studies of the Americas* (pp. 14–44). Palgrave Macmillan. https://doi.org/10.1057/9780230511729_2

Miller, C. (2022, October 5). *The Chips That Make Taiwan the Center of the World*. TIME. https://time.com/6219318/tsmc-taiwan-the-center-of-the-world/

Morgenthau, H. (1948). *Politics among Nations* (1sr. (Digital)). A. A. Knoff.

Morgenthau, H. (1955). Politics among Nations: The Struggle for Power and Peace. By Hans J. Morgenthau. (New York: Alfred A. Knopf, 1954. 2nd ed. pp. xxi, 600, xxv. $5.75.). *American Political Science Review*, *49*(2), 586–586. https://doi.org/10.1017/S0003055400275254

Morillo Moreno, M. C. (2007). Venezuela en el comercio internacional y frente al desarrollo sustentable. *Revista de Ciencias Sociales*, *13*(1), 23–46. https://ve.scielo.org/scielo.php?script=sci_abstract&pid=S1315-95182007000100003&lng=es&nrm=iso&tlng=es

Nakayama, S. (2009a). Chapter 11: Science and Technology in Modern Japanese Development. In *The Orientation of Science and Technology* (pp. 114–136). Brill. https://brill.com/display/book/9789004213074/Bej.9781905246724.i-390_012.xml

Nakayama, S. (2009b). Chapter 25: Eighteenth-Century Science: Japan. In *The Orientation of Science and Technology* (pp. 337–354). Brill. https://brill.com/display/book/9789004213074/Bej.9781905246724.i-390_026.xml

Ohno, K. (2019). Meiji Japan: Progressive Learning of Western Technology. In A. Oqubay, & K. Ohno (Eds.), *How Nations Learn* (1st ed., pp. 85–106). Oxford University Press. https://doi.org/10.1093/oso/9780198841760.003.0005

Ortega y Gasset, J. (1964). Meditación de la técnica y otros ensayos sobre ciencia y filosofía. In *Obras Completas: Vol. V (Sixth)*. Revista Occidente. https://dialnet.unirioja.es/servlet/libro?codigo=156966

Osorio, A. (2022). Why Chuño Matters: Rethinking the History of Technology in Latin America. *Technology and Culture*, *63*(3), 808–829. https://muse.jhu.edu/pub/1/article/859725

Ouyang, H. S. (2006). Agency Problem, Institutions, and Technology Policy: Explaining Taiwan's Semiconductor Industry Development. *Research Policy*, *35*(9), 1314–1328. https://doi.org/10.1016/j.respol.2006.04.013

Parry, R. (2024, Spring). *Episteme* and *Techne*. In E. N. Zalta, & U. Nodelman (Eds.), *The Stanford Encyclopedia of Philosophy*. Metaphysics Research Lab, Stanford University. https://plato.stanford.edu/archives/spr2024/entries/episteme-techne/

Perlo-Freeman, S. (2023, December 5). *Arms Race | Examples, Consequences, & Models*. Britannica. https://www.britannica.com/topic/arms-race

Prebisch, R. (1950). *The Economic Development of Latin America and Its Principal Problems*. https://hdl.handle.net/11362/29973

Rammert, W. (2007). *Technik — Handeln — Wissen: Zu einer pragmatistischen Technik- und Sozialtheorie*. VS Verlag für Sozialwissenschaften Wiesbaden. https://doi.org/10.1007/978-3-531-90485-6

Reese, B. (2018). *The Fourth Age: Smart Robots, Conscious Computers, and the Future of Humanity*. Atria Books.

Reyhani, R. (2007). The Legality of the Use of White Phosphorus by the United States Military During the 2004 Fallujah Assault. *10. JL & Soc. Change*, *10*, 1.

Available at: https://scholarship.law.upenn.edu/jlasc/vol10/iss1/2

Saliu, H. (2020). The Evolution of the Concept of Public Diplomacy from the Perspective of Communication Stakeholders. *Medijska Istraživanja*, *26*(1), 69–86. https://doi.org/10.22572/mi.26.1.4

Santana, F. R. (2017). El desarrollo del comercio internacional del puerto de Punta Arenas (Chile), 1905–1914. *Magallania (Punta Arenas)*, *45*(1), 35–46. https://doi.org/10.4067/S0718-22442017000100035

Teitel, S. (1981). Creation of Technology within Latin America. *The ANNALS of the American Academy of Political and Social Science*, *458*(1), 136–150. https://doi.org/10.1177/000271628145800111

UN General Assembly (26th sess.: 1971). (1972). *Restoration of the Lawful Rights of the People's Republic of China in the United Nations.* https://digitallibrary.un.org/record/192054

Varas, P. A. (1999). Algunas fuentes históricas de la política exterior de Chile. *Estudios Internacionales*, *32*(126), 3–39. https://www.jstor.org/stable/41391622

Wilner, G. M. (1981). The Transfer of Technology to Latin America Symposium: Transnational Technology Transfer: Current Problems and Solutions for the Corporate Practitioner. *Vanderbilt Journal of Transnational Law*, *14*(2), 269–280. https://heinonline.org/HOL/P?h=hein.journals/vantl14&i=287

Wong, G. H. C. (1963). China's Opposition to Western Science during Late Ming and Early Ch'ing. *Isis*, *54*(1), 29–49. https://www.jstor.org/stable/228727

World Bank Open Data. (n.d.). *World Bank Open Data.* Retrieved February 17, 2024, from https://data.worldbank.org

Ye Qing: *Promote the Organic Integration of the Party's Leadership System and the Private Enterprise Governance System-All-China Federation of Industry and Commerce*. (2020). https://www.acfic.org.cn/bhjj/ldzc/hzfhz/yq/yqgzhd/202009/t20200917_60940.html

Yuan, B. J. C., Chang, C.-Y., & Lo, M. C. (1998). Strategies of Semiconductor Industry in Taiwan. *IEMC '98 Proceedings: International Conference on Engineering and Technology Management. Pioneering New Technologies: Management Issues and Challenges in the Third Millennium (Cat. No.98CH36266)*, 541–545. https://doi.org/10.1109/IEMC.1998.727820

3 Latin America

The Epicenter of the Rivalry between the United States and China

Latin America is often treated as a homogenous bloc for policymaking or academic analysis. On the one hand, one could map out the general history of most nations since Spain settled colonies from north to south in the 16th century. During the Spanish colonial period, the Crown imposed a system of resource extraction, regional governance, and policy enforcement, resulting in adopting a standard set of rules, institutions, and regulations across the territories under its control. For example, "*mestizaje*," or Intermarriages between Spanish and native peoples, were common in both countries, unlike in the US or Canada. Latin American countries also gained independence at a similar time and for similar political reasons: mainly through rebel insurgencies against the Spanish crown—conquered by Napoleon at the time[1]—and inspired by the French and American revolutions. Initially, the border configuration followed the traditional lines drawn by the Spanish, but later, it was shaped by local needs, wars, and negotiations. Brazil was always an exception since it gained its independence from Portugal much more peacefully and didn't share a common language with the former Spanish colonies. Two things it does have in common with other Latin American nations—which is essential for this book—are that it is similar in its multicultural demographics and its reliance on natural resources.

On the other hand, Latin America is also profoundly heterogeneous. The region possesses a wide range of climates, geography, and demographics—while varying to some extent in its economic and political models. Cuba is a communist country with a government that has stayed in power for over 70 years, Chile is often considered a *neoliberal* and high-income country, Venezuela struggles with a failed democracy and poverty, and Colombia is regarded for both its consolidated democracy and its widespread violence. In that situation, it is difficult to openly state that any phenomenon that affects the region will have equal effects on each state.

This chapter is broken down into concise sections that provide an overview of the context and conditions under which the US and China compete for influence and reputation in Latin America. It also explores how this influence impacts the discussion around technology implementation and the remaining gaps in analysis. A final objective for this chapter is to clarify why it is natural to consider the US' position in security discussions related to Latin America and China's increasing presence.

DOI: 10.4324/9781003489450-4

Latin American Historical Influence from the US

The US has primarily influenced Latin America, and there are institutional and historical frameworks that have contributed to this relationship. The fact that the American Revolution inspired many independence processes for Spanish colonies against the English and the same Enlightenment ideas is arguably a political and foundational link between the North and the South. The US has pushed its influence in the region for centuries and has contributed to political processes, democratization, military defense, fragmentation of countries, and institutional development for individual states and the region.

The Monroe Doctrine of 1823 allowed the US to intervene in the region to protect its interests and shape foreign policy. Under this premise, the US directly supported the Cuban and Puerto Rican independence from Spain, funded the construction of the Panama Canal—during its independence from Colombia, and intervened in several nations in Central America during the interwar period. Later, during the Cold War, the US intervened in several anti-communist movements, supported operations through the American Central Intelligence Agency (CIA), and partially ignored the loss of democracy when it served its purposes. Despite all the former, the US influence in the region has also been locally interpreted as positive regarding economic, military, and political support, as well as a defense against external threats and democratization.[2] Consequently, "US policy in the hemisphere had always emphasized the shared bond of colonialism, independence and commitment to freedom and liberty with its Latin American neighbors."[3]

China's Historical Connection with Latin America

In spite of the geographic and cultural distance between China and the Americas, the PRC's approach to this region is not random—China has a long history of interaction and diplomatic relations across the Pacific. Records point to trade with the Spanish Empire through the Manilla Galleons, and there was communication between the former colonies and the Qing dynasty, and later with the importation of "*coolies*" or Chinese migrant workers, many of whom have integrated into societies and shared their culture. One of these destinations, Peru, was also the first country to establish formal relations with China in 1875.[4] Latin American countries were also among the first to recognize the Chinese Communist Party (CCP) as the legitimate government of China. Cuba established relations in 1960, a natural decision considering that both countries shared ideological frameworks. In December of 1970, two years before President Nixon visited the PRC, the socialist President of Chile, Salvador Allende, established direct diplomatic relations with Beijing. Even after General Augusto Pinochet, an anti-communist leader, seized power in 1973, this policy continued to be enforced.[5] Consequently, Chile was also among the first countries to establish a Free Trade Agreement with the PRC, signed in 2005 and enacted in 2006.[6]

China's interest in the region seems to be economic—as a source of natural resources—and political. Its diplomatic objective openly contrast the US dominance in the neighborhood, and also aim to reduce Taiwan's participation internationally, since central and South America were among the nations that kept Taipei's recognition. China invested resources and pressured numerous governments to put an end to this trend by convincing Central American countries such as Costa Rica and Panama to cut off relations with Taipei in favor of Beijing, something that in most cases did not pay off.[7] As for 2024, only a few countries recognize Taiwan in the region, including Guatemala, Haiti, Belize, and Paraguay.

Geopolitics and Domestic Security in the Region

As expected, historical background and geopolitical context heavily influence state-to-state relationships in Latin America. The region has experienced fewer interstate wars than others. Conflicts since the late 1800s have been scarce, and though there are border and territorial disputes, these have been kept in diplomatic negotiations since the end of the Cold War.[8] Recent tensions, such as the Venezuelan reclamation of western Guyana[9] or the Argentinian and Chilean sovereignty over Antarctica,[10] have been kept in the diplomatic sphere.[11]

Security challenges are, therefore, different from those in other regions. According to a yearly report by the Centro de Estudios Internacionales UC (CEIUC), the region's most urgent political and geopolitical risks are mostly related to domestic conflicts, uncontrolled migration, corruption, and international crime. Nevertheless, international conflicts are constantly getting more attention. In their 2024 issue, they point out that domestic security and organized crime have risen as homicides increased and narcotraffic became more prominent,[12] even in areas where it was relatively controlled, like Ecuador. On top of that, there is general dissatisfaction with the judiciary, democracy, and corruption, which results in a more skeptical population that is apathetic toward politics.[13] The 2023 Latinobarometer survey [Latinobarómetro] reports that 28% of Latin Americans do not care if their country is democratic—the highest percentage on record. About 48% consider democracy the preferable form of government, but this is low considering that 63% supported democracy in 2010. The adjacent report states that it can be correlated with the diminished economic development the region has undergone since the early 2010s.[14] I concur with this takeaway and argue that a less democratic population might also be more vulnerable to authoritarian influence since there is a lower appreciation of democracy as the preferable form of government.

Among all those regional and local issues, the CEIUC report dedicates two subtitles to more globalized affairs. On the one hand, cybersecurity and digital risks, which the region is vulnerable to, are highlighted. On the other hand, international instability is worrisome. The authors highlight issues concerning the US–China rivalry and escalating conflicts, impacting Latin America's economy and political integrity.[15]

International Latin American Order

International organizations significantly influence Latin America's foreign and regional policies. Although the region has never achieved the level of integration seen in the European Union, there are initiatives that either embrace or reject US power in the hemisphere. These initiatives impact policymaking and geopolitical alliances in the region, and this book is also a concern. The Organization of the American States (OAS) was founded in 1948 to promote the idea of Pan-Americanism, continental security, democratization, the prevention of inter-state war, economic and social development, and discuss any normative and legal framework that concerns the region.[16] The OAS has, at the same time, partly tolerated authoritarianism. While Cuba was excluded from the organization in 1962, right-wing authoritarianism was generally more allowed.[17]

Although many realists believe this form of organization reflects the US hegemony over the region, the OAS has had operational and political achievements, as well as normative framework outputs that serve as proof of its role in limiting the potential negative impact of regional instability and defending democracy, at least until the early 2000s.[18] In technology and security, the OAS has pushed countries to increase their cybersecurity capabilities for over 20 years, with mixed results. The CICTE's Cybersecurity Program intends for governments to enact policies, involve stakeholders, and promote intelligence to face increasing cybersecurity risks and raise awareness. Of course, the program's strategic partners reveal a particular set of values and a vision related to the definition and means to answer cybersecurity threats. For example, the Canadian, UK, Spanish, Estonian, and US Governments are listed as strategic partners, as well as the Global Forum on Cyber Expertise (GFCE), The World Economic Forum, Oxford University, Florida International University, Microsoft, CISCO, Amazon, Facebook, City Bank, and X (Formerly Twitter).[19]

However, the loss of democracy in the region and the rise of critical coalitions to the US has led to significant changes in political systems and overall governance. The Forum of Sao Paulo first met in the 1990s to provide an "anti-imperialist" agenda and challenge economic and political integration with the US by proposing a Marxist alternative or cooperation of left-leaning governments. Also, to rely on regional alternatives like MERCOSUR and CAN.[20] A series of initiatives follow a similar line of thought, although they often depend on the incumbent government ideology. Moreover, it has been within these alternative regional organizations that authoritarian governments have found ideological support. This is the case with Venezuela, Cuba, and Nicaragua. The Economic Commission for Latin America and the Caribbean (CELAC in Spanish), which does not include the US or Canada, is a more recent endeavor that has the aim to become an international organization but—as with the Sao Paulo Forum—still lacks the continuity and formality the OAS has,[21] even though it has become a door of influence for the PRC since the China–CELAC forum is held—ideally—once a year to enhance the region's integration with Beijing.

The US as a Hegemon and China as a Challenger

For many, the end of the Cold War was a turning point for the US, solidifying its dominant international position. For decades, the US led a Western military alliance and emerged as the sole superpower after the collapse of the Soviet Union.

Even with its military strength and extensive capabilities, the US was not seen as a threat by its European allies post-Cold War, particularly within the framework of NATO. This was because of a mutual understanding of values and ideology.[22] After the USSR dissolved, the emerging international order prioritized cooperation and mutual understanding, with the US seen as a significant power. The focus was on the benefits of trade and collaboration rather than the possibility of an ongoing conflict that could escalate into war.

Moreover, at that time, there weren't any alternative states that could counter Washington. Russia was undergoing significant political transformations, while China was still in the early stages of its more *open* economic policies. Consequently, analysis shifted to potential multipolar or non-traditional conflicts, such as the War against terrorism, global financial crises, or the role of international institutions in setting up rules and frameworks for state coexistence.[23] Nevertheless, in recent years, the US' presence has arguably diminished, and the West, in general, has been said to be "retreating" from its internationally dominant podium. This is for several reasons. For example, anti-globalist positions have increased among voters[24] and have fed narratives that reject the ideals of a liberal order that prioritizes cooperation over sovereignty.

On the other hand, the US has continuously retreated from its realm of influence. During the later years of the Cold War, the US shifted from prioritizing military aid programs to economic cooperation as a tool of influence,[25] which is still a strategy used today but has considerably diminished in response to protectionist measures. In Latin America, the loss of interest and rise of alternative forms of influence followed Washington's policy after 9/11.[26] In this scenario, instability, divisions, and economic transformations increased.[27]

While the US is undergoing essential transformations in its international policy, defiant actors have also transformed their position and attitude toward the international order. Russia is aggressively expanding its territorial claims and influence into other countries around the world under Putin's leadership. Similarly, the PRC was, at least during the 1990s and early 2000s, on its way to being a part of the international system, following rules and promising internal reforms. States looked at Beijing skeptically but with the hope that the liberal order and the third wave of democratization could also lead to openness in their system.[28] Education, economic, and living conditions rose, and in 2001—after 15 years of negotiations—China joined the World Trade Organization (WTO) under certain conditions,[29] which it has not fully complied with. Since the late 1970s, Beijing has pushed for reforms to its economic and political system as a way to avoid the fate that had befallen the USSR.[30] Since the arrival of Xi Jinping into power, the state has become increasingly more authoritarian and moved further away from the technocratic system of the reform era.[31] China's foreign objectives have also changed, now better armed to challenge the international order and to mold it to their interests. Technological expansionism is, therefore, one of many tools they might use to increase their presence abroad. Different from Western nations, in which innovators are independent actors, the PRC holds the advantage of being able to utilize their private assets abroad. Companies are often institutionally pressured to follow the PRC's international strategies,[32] and overseas Chinese individuals are also subjected to Beijing's guidance to achieve objectives in the international theater.[33]

US–China Competition across the World

In 1971, Henry Kissinger visited China, a milestone that marked the beginning of a tense yet pragmatic relationship between both states. In his 2011 book *On China*, Kissinger reviews the country's history, diplomatic tradition, and Beijing's inevitable rise.[34] He also compares the US and Chinese positions in the international sphere, opening questions of how sustainable they are in the long term. These questions have taken new significance recently as China has developed its technological, military, and economic capabilities.

Therefore, the relationship between the US and China differs from that at the beginning of the century. The current convention is that China will inevitably rise and challenge the global order and that the US is reacting to this as an existential threat to the world's hegemonic power.[35] As early as 2001, Mearsheimer argued that the US should adopt a balancing policy against China to prevent it from becoming too powerful.[36] In that sense, essential provisions have been raised regarding the potential trade opportunities and future economic conflicts between the two countries.[37]

It is not a secret that, in recent years, the US and China have experienced increasing tensions, mainly since the Donald Trump administration's narrative made it explicitly relevant to their government program—even when security concerns were raised a few years before that. In 2018, tangible measures were taken when economic tariffs were imposed on both sides due to national security, intellectual property issues, unlawful competition, and other causes.[38] Political tensions have spread to various domains, including Taiwan, overseas investments, cinema culture, space exploration, and global health.

Technological Competition and Political Objectives

This competition inevitably includes technology, an area where the US and the PRC discussion is not only a narrative one, but it has propelled actual policy decisions and international pressures that indirectly affect countries located in the middle.[39]

An arrangement of researchers has explored the confrontation in telecommunications, that is, the technology that allows for interconnectivity. They often highlight these two actors as the leading players in creation, dissemination, and rule-setting around the globe.[40] The US advocates for security and like-mindedness, while the PRC attempts to gain territory with aggressive yet effective market strategies through its companies.

The advantages and hindering factors underneath each actor have also been widely explored. For instance, some claim that Chinese efforts to dominate the international telecommunications market have not been as successful as intended and expected, as decision-makers worldwide have questioned companies such as

Huawei or ZTE due to potential security risks.[41] Australia began assessing risks in the telecommunications sector at the beginning of the century and made proactive efforts to increase security in this sector in 2018. Although it wasn't explicitly directed to the PRC, Chinese companies were the most affected by the results.[42] Nevertheless, some propose that the Chinese state has the upper hand over the American one because of the latter's excessive bureaucracy and lack of supply and that the capacity to create, own patients, and disseminate technologies is a vital issue.[43] In other cases, experts have argued that decision-makers overlook many security risks if they lead to more flexible development for their countries.

The Latin American Question

In Latin America, which, as mentioned before, is deeply connected to the US and has increased influence from China, a similar assumption has been raised related to a potential geopolitical struggle influencing decision-making when implementing new technologies. Latin American states seek development but are concerned about cybersecurity and possible threats. Balancing these two needs directly correlates with how they respond to potential technology providers and the speed and effectiveness they apply to their systems. To gain a more comprehensive picture of the extent and nature of the presence of the US and China in Latin America, I compare economic relations, the reputation of both nations among the population, and the questions related to the provision of technology and ongoing debates on potential security concerns. When reviewing specific cases in later sections of this book, I provide further context for each situation and a more particular analysis of influence, coercion, and lack of transparency associated with those scenarios.

The US might have partially retreated from the region, but its economic presence is still strong. ECLAC states that around 20% of the US Foreign Direct Investment (FDI) went to Latin America in 2014, with holding companies representing 44% of those transfers. The main countries in South America that US investment was directed to that year were Brazil, Chile, and Venezuela.[44] According to the US Bureau of Economic Analysis (BEA), in 2021, global FDI was US$6,489,012 million, and the state's total accounted investments to South America reached US$141,099 million, while the total Latin America and Western investments represented 15.7% of the country FDI at US$1,017,716 million.[45]

According to the same institution, technology-related investments only amount to a relatively small percentage. For example, computers and electronics represent 2.9% of the US FDI in South America, the same as professional, scientific, and technical services. While "information" accounted for a larger share at 6.3%.[46] Nevertheless, the definition of "information" investments is quite ambiguous and challenging to narrow down, as established by the "North American Industry Classification System (NAICS)":

> The Information sector comprises establishments engaged in the following processes: (a) producing and distributing information and cultural products, (b) providing the means to transmit or distribute these products as well as

> data or communications, and (c) processing data. The main components of this sector are motion picture and sound recording industries; publishing industries, including software publishing; broadcasting and content providers; telecommunications industries; computing infrastructure providers, data processing, Web hosting, and related services; and Web search portals, libraries, archives, and other information services.[47]

The US is also the primary or second major economic partner of most Latin American countries. This is strongest in the cases of Mexico and Central America, despite the increase of Chinese imports to the region, which is constantly highlighted by decision-makers.[48] Mexico is the second-largest destination of US goods and was the second-largest source of US imports in 2022. For the Mexican part, the US is undoubtedly irreplaceable in the foreseeable future, since—according to the Observatory of Economic Complexity (OEC)—during 2022, 76.8% of its exports landed in the US, while 55.6% of its imports came from its northern neighbor.[49] The Central American case is similar. For instance, the US continues to be Costa Rica's leading trading partner and investor, despite Beijing's promise to boost investments and trade if Costa Rica abandoned Taipei.[50]

The PRC's most substantial economic influence area in Latin America is the volume of bilateral trade, which is particularly strong in South America. A relatively new phenomenon, the exchange between China and the region has risen from $12 billion in 2000 to over $430 billion by 2021. In South America, it is the largest trading partner of Argentina, Brazil, Chile, Paraguay—even though they don't have diplomatic relations—and Uruguay, while in Central and North America, it is still behind the US as a trading partner.[51] China's economic trade mainly imports natural resources and exports machinery and technological products, making it a natural player when considering the geopolitics of technology transfer. Its economic presence in the region's technology sector is expected to be reflected in the type of foreign investments the country allocates. However, the PRC's investments primarily concentrate on infrastructure, energy, and natural resources-related activities.

The OEC also provides a picture of China's exchanges with the region. It showcases that its export grid does not depend too much on Latin America but on the US, its leading destination, followed by Japan, South Korea, and Vietnam. The PRC primarily imports from Australia, South Korea, Taiwan, and the US.[52] In that sense, changes in trade in Latin America might be easily overlooked. Nevertheless, I argue that its fastest-growing export markets are the US, Brazil, and México, which indicates a complex development in the region. Since Mexico's imports from China have risen more than 60% in 2024,[53] it is possible to infer that the decrease in American demand has caused Beijing to divert exports.

As for its counterparts, if the US's influence on trade is more significant in North and Central America, the PRC is almost indispensable in South America. Chile's economy is most affected by these relations, making it perhaps the most problematic case to study in the long term despite its strong institutions. Exports to China represent around 40% of everything the nation sends abroad, representing around 12% of Chile's GDP. In this scenario, several warnings have been raised on how the country can balance its relationship with the US—as its closer ally—and China's growing grip. Chile is followed by Peru and Brazil, where exports to the PRC represent 10% and 7% of their respective GDPs.[54] Imports from China are also snowballing, especially since a series of agreements and benefits make transport and customs operations cheaper and swifter. In other words, there is evidence of an implicit influence that China exerts over these countries, which has been incentivized by this multi-factorial dependence.[55]

China is also present in the so-called "Triangle of Lithium," a geographic location extending between Chile, Argentina, and Bolivia, accounting for 52% of the world's reserves and a third of the current production. As the world is becoming more dependent on batteries that use this resource, the PRC has extended its strategy to own as much as possible of the lithium supply chain, including the ownership of two of the largest lithium companies, Ganfeng and Tianqi, as well as purchasing a significant share of Chile's SQM in 2018.[56]

The Belt and Road Initiative is another way China has increased its regional presence. By the end of 2023, only Paraguay, Colombia, and Brazil had remained out of the South American initiative, while its membership in Central America was almost unanimous.[57] Moreover, Mexico has not joined yet but has been mentioned as a target partner. Many PRC investments, loans, and projects are embedded in this framework.[58] Although some projects have sparked issues with the local population and with trust toward China, social and political needs often overtake the doubts due to concerning behavior from the PRC counterpart. In the case of the Coca-Codo Sinclair dam in Ecuador, it has been observed that the project has resulted in a considerable amount of debt owed by the country to Chinese institutions. Moreover, the project has exacerbated social and environmental risks.[59]

A 2021 ECLAC report shows that FDI in Latin America fell considerably during 2020, mainly due to the COVID-19 pandemic and the international crisis.[60] That same report indicates that China topped the first and third positions in regional investments in 2020 when Chinese power companies acquired American Sempra operations in Peru and Chile.[61] However, the European region has been the leading investor in the region, contributing around 50% of FDI inflows from 2010 to 2019 and 38% in 2020. The US accounted for 27% of investments between 2015 and 2019 but increased its share to 37% in 2020. On the other hand, China's contribution was less than 5% and is grouped under "others," while Japanese investments accounted for 4% in 2020.[62]

In summary, both countries strongly influence Latin America's economy, but their areas of focus and strategies differ from one country to another. China is interested in Latin America and the Caribbean, and its foreign policy reflects this. Between 2008 and 2021, this target region received 26% of Chinese loans, making it the second largest destination after Asia, which received 36%.[63] In 2021, Chinese FDI in Latin America was the second highest behind Europe, at US$11.8 billion.[64] Chinese investments are also promoted by international organizations, such as the United Nations Economic Commission for Latin America and the Caribbean (ECLAC), and the Belt and Road Initiative has even been addressed as a "civilizing connectivity proposal for shared prosperity," showcasing an overall trust in localized operations.[65]

The US operates more through FDI, whereas China has pushed for more bilateral trade, importing natural resources and exporting their finalized products. Nevertheless, significant projects from China, some more sensitive or strategic than others, have sparked concern in the US.[66] Examples of concerning investments include the installation of military bases disguised as space exploration centers—which I cover in a later chapter—investments to dominate critical markets such as the lithium supply present in the Chile-Bolivia-Argentina triangle, as well as investments in technology that deals with sensitive information, and the energy sector. There are mixed interpretations on whether certain behaviors from China to increase its presence and grip throughout the region are corrosive to institutions and democracy, especially when an investment or scenario involves state-to-state agreements, economic statecraft, and traditional corruption.[67] Analysis shows that China has entered markets and social environments—with loans, FDI, trade, and so forth—by taking advantage of the US retreat in the region. In other words, the PRC acts opportunistically when it resorts to economic statecraft through government policy.[68]

Reputation and Political Influence

What is the public image of both these competing states? While producing this book, I interviewed key actors from the region, including decision-makers, experts, and civilians, to gather their opinions on China and the US. These conversations provided me with an overall understanding of Latin Americans' skepticism toward both nations, particularly toward Beijing, following the COVID-19 pandemic.

Some research endeavors have attempted to measure states' influence and soft power in Latin America outside of traditional forms of coercion, with mixed results. One recent study by Urdinez and Winters outlines the perception of a set of countries toward international powers. Although the study intends to show the effects of China's public diplomacy surrounding COVID-19, it surveyed a thousand cases per country to compare the perceptions of a selection of international powers, including the US, the PRC, Japan, France, and Italy, as controls. The results showed that China could not improve its public image to the extent it was expected. Keywords such as "virus," "wall," "technology," "communism," and "cheap" were among the most mentioned when surveyed individuals talked about China.[69] The US, on the other hand, was related to "power," "money," "capitalism," and "Trump," among others. Interestingly, Japan was the country most associated with technology among the sample.[70] That is, the word people mentioned first when thinking about the target country.

When strictly comparing public opinion on both nations, data from the Pew Research Center between 2014 and 2017 portrays that the US experienced a drastic decline in its favorable perception among the people of Latin America. In Chile, it fell from 72% of support down to 39%. There is a more extensive division in Colombia, Peru, Bolivia, and even Venezuela, as the most recent values range from 55% to 47% of favorable views toward the US.[71] Nevertheless, 2023 data from the Latinobarometer show that these views have recently changed. Of the 19,205 respondents surveyed, 71.1% indicated they have a positive or generally positive image of the US. The lowest are Argentina and Chile, with around 56% support—and Venezuela, with 61% of people declaring they have a positive image of the US.[72] Several reasons could explain this improvement in regional support, including the US' conscious return to the international sphere or historical reasons.

Since Xi Jinping took power, China's overall perception has shifted from positive to negative, according to The Pew Research Center.[73] Nevertheless, Africa and Latin America remain friendlier toward China than the European and American West. According to the same center, 41% of Argentinians, 39% of Brazilians, and 57% of Mexicans had a positive view of the PRC in 2023, way under the median of 28% of positive perception.[74] In other words, the population's skepticism toward both countries remains polarized. Still, the PRC has opportunities to gain support, considering the US loss of trust and the economic crisis Latin America has faced since the late 2010s. Based on the Latinobarometer survey, 51% of Latin Americans consider China to be very trustworthy or somewhat trustworthy, while 41.5% believe that it is barely or untrustworthy. The opinions on whether China interferes in local affairs are more split, with 41% believing it does and 48.7% believing it doesn't or only slightly interferes. Finally, 37.6% consider China's model to be *good* for their country to follow, while 52.3% refuse this statement.[75] In other words, while China's image is evidently better in Latin America than in other regions, there is division on how this country should be addressed.

At this point, I find it essential to highlight that public opinion toward these two nations is relevant, as decision-makers from democratic countries in the region consider public pressure when evaluating different options for a project. However, the opinion of the elites—and the cooption of said groups—in Latin America is disproportionally vital in the outcomes of these processes. In technology transfers, the US and China directly negotiate, work, and persuade key decision-makers or prominent figures in the private and public sectors. Therefore, although I consider the general public's perception, the analysis shows that the interaction of narratives, perceptions, and technical assessments funnel into the decisions of technology transfers.

Rivalry on Research, Innovation, and Geopolitics

The rivalry between the US and the PRC in terms of influence in Latin America transpires to many areas. The US is close to several academic institutions in academia due to the reputation of its universities and the *Fulbright* program.[76] At the same time, China has invested considerable resources to promote and extend its academic public diplomacy.[77] Efforts by China are focused on enabling the transfer of technology to Latin America through two main channels—the Belt and Road Initiative and Huawei. With its strong regional presence, Huawei provides academic support, technological expertise, and tangible hardware to facilitate this transfer.[78] The US, on the other hand, cannot direct the actions of its private companies the way China does, so its actions are limited to the political and diplomatic arena.

Some researchers in the field argue that the PRC has positioned itself as a direct competitor to the US on emerging technologies and that the PRC has increased its prestige in the region,[79] hence its soft power capabilities. The implication of this comparison—and the general framing of the issue—is that states are meant to choose one of these two states' options. China has gained ground because its offer is more affordable than the competition.[80]

> Throughout this process, Latin America faces a double challenge: (1) technological restraint resulting from a peripheral position in the creation and management of technological flows and resources and (2) growing pressure from preeminent global players, such as China and the United States, so that countries in the region adopt technological standards and systems in the area of telecommunications, produced (generated) by public or private corporations that respond to both contenders.[81]

The confrontation between the US and the PRC has been central to the discussion of geopolitics and technology dissemination. I criticize that this attractive, yet narrow vision has led some authors to dismiss any other interaction with the region, but only with these two countries. Contrasting Washington and Beijing as two main technological forces in the area—an important issue to debate—must be considered carefully.

Diverse entities not directly connected with the US or the PRC serve Latin America's technology sector. Some key players in this field are private companies from Europe, Korea, and Japan. These market competitors have demonstrated a strong track record of delivering cutting-edge technological solutions to countries throughout the region. I argue that the role of middle and regional powers must be considered in this debate for all these factors.

In the discourse surrounding technology and innovation, some providers—that are less contentious—are frequently overlooked by the media and scholars, who tend to concentrate their analysis on the US and the PRC. In some cases, American companies don't even bet on these processes, like in the race for 5G connectivity between and within regional states.[82] In other cases, an intricated and coordinated effort often involves several companies and nations in the process. In 5G public tenders, for example, the operators that present their offers to a state entity differ from the manufacturers that provide the technology. Even though some states acknowledge these relationships beforehand, in other processes, the partner company is not disclosed or barely mentioned when they present their offers. When understanding public discussions, decision-making processes, and the outcomes of nationwide technology implementations, it is essential to consider all these factors and understand information availability. In this scenario, entities such as Nokia and Ericsson and development organizations like the Inter-American Development Bank or the OECD must be considered actors outside the US and China rivalry.[83] Europe can be viewed as a significant bloc of influence in the technology competition, as the whole EU is a normative leader and economically relevant to the region, even compared to China and the US.[84]

This complex interaction of actors applies to other sensitive or critical technologies, such as facial recognition software—now being discussed in the region—[85] optic fiber standards for connectivity, and many forms of keeping and managing data.[86]

Middle and Regional Powers Complementing and Challenging the US and the PRC

If Chinese and American companies are not in direct opposition, it is essential to explain the role of these interacting states and private actors in the overall context of power and influence competition in the region. In Latin American regional and international policy, there are also regional powers and *leading*[87] states that push for changes or that are used as models for their neighbors.[88] But where do these powers lie in decision-making and international relations? *Mittelmacht, or middle power,* is historically connected to the German tradition and traditionally referred more to location than quantitative capacity and power. It refers to the states that established their sovereignty between Russia and France, such as those that had enough or relative power to maintain their independence. Nevertheless, this idea has changed over time, as countries such as Canada and Australia began to identify themselves as "middle powers" after the Second World War.[89] As the notion developed, it became more interconnected with the conceptualization of geopolitical hierarchy[90] and less to the idea of independence since it does not guarantee significant power within the international system.[91]

But not all middle powers are homogenous. Traditional middle powers are most wealthy and stable but often don't have regional power, while modern middle powers are associated with young democracies wielding economic growth.[92] Professor Jeffrey Jordaan explains that:

> Middle-power foreign policy often focuses on conflict reduction (broadly understood) in the world system by involving other like-minded states (in terms of the issue at hand) in an attempt to arrive at a workable compromise, usually through multilateral channels and institutions. Despite some disagreements with the hegemon and other major states (especially over human rights issues), **middle powers do not challenge or threaten the global status quo** —that is, the economic and military to political 'balance' of power— nor the desirability of liberal democracy, in any fundamental way.[93]

Middle powers are generally more focused on cooperation and collaboration over confrontation, and they share behavioral characteristics in how they address international relations, such as the use of diplomacy and sometimes mediation to achieve practical changes—for authors such as Chapnick, middle powers are defined not by their resources but by how they interact in the international theater.[94]

According to these definitions, countries such as Japan, Sweden, or Brazil can be defined as middle powers.[95] However, it depends on the definition used to frame such powers. Sometimes, like in the case of Brazil, the characterization requires "stretching the concept," leading to misconceptions.[96] A series of tools can be applied to the analysis to narrow down this description. For example, Professor Andrew Carr identifies a series of categories of middle powers depending on their position, their extension, behavior, or how they exert diplomacy, and their identity as if the country identifies as such "middle power" or doesn't.[97]

Neoclassical realism is a helpful framework for understanding this role, as it considers other factors that make more minor powers competitive.[98] It has been addressed to connect the actions and decision-making of middle powers with China's influence abroad, but results widely vary. For example, a paper by Özşahin et al. elaborates on Turkey's reasoning behind its interactions with the PRC and the US as a middle power and how its internal affairs are intertwined with its international behavior.[99] In the case of the Japan–China relationship, historical events, cultural factors, the presence of nationalism, and domestic affairs are also connected to both states' external policies beyond traditional notions of power.[100] Overall, the actions of the middle powers seem to reflect domestic and cultural traits in their international decision-making processes, and I consider this to be coherent with neoclassical realism.

In the context of technology adaptation, middle powers might be providers and receptors of the innovation frameworks in question. This issue will depend on these actors' role in the international system and how they identify. A clear case of this is Korea, which has long sought to be recognized as such middle power,[101] and that is also the home country of technology companies that are studying and deploying telecommunications in the South American region.[102]

In this chapter, I discussed the intricate nature of comparing the strategies of the US and the PRC due to Beijing's stronger grip over its companies' conduct, bolstered by both domestic and international legislation. In contrast, Washington primarily relies on incentives and indirect political or economic influence, with few but notable examples of direct pressure. This issue is not unique to the US but extends to other democracies, which confront limitations in steering their strategies pertaining to their private sectors.

When reviewing the dynamics of US–PRC influence and competition in Latin America, particularly regarding the transfer of technology, the concept of a potential new Cold War bifurcating the world into ideological or value-driven blocs may be illuminating. From this vantage point, it could be posited that opting for an alternative to the Chinese or American model does not negate the existence of an ideological contest over economic and political frameworks, one that reverberates across both the influencing and influenced countries. In light of this, a significant portion of Latin America's technology providers exhibit closer political alignment with the US than with China, hinting toward a geopolitical positioning that transcends mere transactional relationships in the domain of technological exchange.

Notes

1 Except for Brazil, which was then ruled by the Portuguese, who allowed for a peaceful independence.
2 Sabatini (2013), pp. 2–4; Schenoni and Mainwaring (2019).

3 Sabatini (2013), p. 5.
4 Creutzfeldt (2016), online.
5 Matta (1991), p. 349.
6 Biblioteca del Congreso Nacional de Chile (2010), online.
7 Rivera-Matias (2022), online.
8 Cairo Carou and Lois (2014), online.
9 Even though Venezuela pressured its neighbor by increasing security in the border.
10 Vidal (2023).
11 Goodwin (2024), online.
12 Sahd et al. (2024), p. 11.
13 Sahd et al. (2024), pp. 15–21.
14 Lagos (2023), p. 19.
15 Sahd et al. (2024), 27–28.
16 OAS (2009), Article 2, online.
17 Segovia (2013), p. 98.
18 Boniface (2002), pp. 376–377.
19 OAS (2003), front page.
20 "Progress at Foro De Sao Paulo," (2002), p. 1.
21 Segovia (2013), p. 101.
22 Press-Barnathan (2006), online.
23 Press-Barnathan (2006), online.
24 Trubowitz and Burgoon (2022).
25 Fitch (1993).
26 Hakim (2006), p. 40.
27 Hakim (2006), p. 42.
28 Minzner (2020), pp. 21–23.
29 Agarwal and Wu (2004), online.
30 Minzner (2020), p. 21.
31 Minzner (2020), p. 179.
32 Tomasic (2015), Wu (2016), p. 265.
33 Hua To (2014).
34 Woo (2011).
35 Nevertheless, others argue that the PRC should not pose a threat or a defiant state against the international order. This is because it does not measure in comparison to the United States—when addressing proximity or aggregated power—and it is rather "perception of aggressive intentions" or "economic considerations" that urge Washington to react (Foulon, 2015, p. 651)
36 Mearsheimer (2001), p. 4.
37 Friedberg (2005); Wu (2016).
38 Bown (2021).
39 Tekir (2020), online.
40 Lee (2020); Tiezzi (2020); Kaska et al. (2019).
41 Umbach (2020), pp. 2–4.
42 Botton and Lee-Makiyama (2018)
43 Lee (2020), online.
44 Comisión Económica para América Latina y el Caribe (2015), online.
45 U.S. Bureau of Economic Analysis (BEA) (2023), online.
46 U.S. Bureau of Economic Analysis (BEA) (2023), online.
47 North American Industry Classification System (NAICS) U.S. Census Bureau (2022), online.
48 Office of the United States Trade Representative (2023), online.
49 Observatory of Economic Complexity (2023), online.
50 Rivera-Matias (2022), online.
51 Polga-Hecimovich (2022), Online.

52 Observatory of Economic Complexity (2022), online.
53 Cota (2024), online.
54 Takahashi (2023), p. 5.
55 *The Belt and Road Initiative in Latin America* (2022)
56 Sanchez-Lopez (2023), online.
57 (Wang, December 2023), online.
58 Myers (2018), p. 241.
59 Vallejo et al. (2019); Gamso and Moffett (2023).
60 Comisión Económica para América Latina y el Caribe (2021), p. 24.
61 Comisión Económica para América Latina y el Caribe (2021), p. 34.
62 Comisión Económica para América Latina y el Caribe (2021), p. 39.
63 Takahashi (2023), p. 2.
64 Takahashi (2023), p. 3.
65 Comisión Económica para América Latina y el Caribe (2017), online.
66 Cheng (2023), online.
67 Shoujun et al. (2018), p. 263.
68 Urdinez et al. (2016), p. 24.
69 Urdinez and Winters (2021), p. 10.
70 Urdinez and Winters (2021), annex, p. 4.
71 Pew Research Center (2023), online.
72 Lagos (2023), question 157.
73 Silver et al. (2022), online.
74 Pew Research Center (2023), online.
75 Lagos (2023), questions 179, 180, 183.
76 Bernasconi (2008), pp. 13, 105–107.
77 Baisotti (2023), pp. 8–9.
78 Baisotti (2023), pp. 13–16.
79 Colombo et al. (2021), pp. 97–98.
80 Balbo and Cesarín (2020), p. 48.
81 Balbo and Cesarín (2020), p. 25.
82 OECD (2019), pp. 41–42.
83 OECD (2019).
84 Moravcsik (2009); Moravcsik (2010).
85 Venturini and Garay (2021).
86 Kobek and Caldera (2016).
87 States that are often referenced and followed despite now being one of overwhelming economic and military power.
88 Neto and Malamud (2015).
89 Robertson (2017), online.
90 Holbraad (1971), p. 78.
91 Holbraad (1971), p. 86.
92 Jordaan (2003), pp. 181–183.
93 Jordaan (2003), p. 168.
94 Chapnick (1999).
95 Holbraad (1971), p. 88.
96 Burges (2013).
97 Carr (2014), pp. 73–76.
98 Jackson (1987).
99 Özşahin et al. (2022), pp. 219–220.
100 King (2013), p. 460.
101 Mo (2016), pp. 588–589.
102 Zaballos et al. (2020)

Bibliography

Agarwal, J., & Wu, T. (2004). China's Entry to WTO: Global Marketing Issues, Impact, and Implications for China. *International Marketing Review*, *21*(3), 279–300. https://doi.org/10.1108/02651330410539620

Baisotti, P. (2023). China's Charm Offensive in Latin America and the Caribbean: A Comprehensive Analysis of China's Strategic Communication Strategy across the Region [Part III: Image, Academia, and Technology]. *Research Publications*. https://digitalcommons.fiu.edu/jgi_research/57

Balbo, G., & Cesarín, S. M. (2020). ¿Guerra comercial, periferia tecnológica o tecnoimperialismo? : América Latina ante la competencia global en el sector de las telecomunicaciones. https://repositorio.aladi.org/handle/20.500.12909/30827

Bernasconi, A. (2008). Is There a Latin American Model of the University? *Comparative Education Review*, *52*(1), 27–52.

Biblioteca del Congreso Nacional de Chile. (2010). *TLC entre Chile y China continúa dando frutos a 4 años de su firma—Programa Asia Pacifico* [Text]. Observatorio Asiapacifico; Biblioteca del Congreso Nacional de Chile. https://www.bcn.cl/observatorio/asiapacifico/noticias/tlc-chile-china-cuarto-aniversario

Blaufarb, R. (2007). The Western Question: The Geopolitics of Latin American Independence. *The American Historical Review*, *112*(3), 742–763. https://doi.org/10.1086/ahr.112.3.742

Boniface, D. S. (2002). Is There a Democratic Norm in the Americas? An Analysis of the Organization of American States. *Global Governance*, *8*(3), 365–381. https://www.jstor.org/stable/27800350

Botton, N., & Lee-Makiyama, H. (2018). 5G and National Security after Australia's Telecom Sector Security Review. *European Centre for International Political Economy (ECIPE), Brussels, ECIPE Policy Brief.*

Bown, C. P. (2021). The US–China Trade War and Phase One Agreement. *Journal of Policy Modeling*, *43*(4), 805–843. https://doi.org/10.1016/j.jpolmod.2021.02.009

Burges, S. (2013). Mistaking Brazil for a Middle Power. *Journal of Iberian and Latin American Research*, *19*(2), 286–302. https://doi.org/10.1080/13260219.2013.853358

Cairo Carou, H., & Lois, M. (2014). Geografía política de las disputas de fronteras: Cambios y continuidades en los discursos geopolíticos en América Latina (1990–2013). *Cuadernos de Geografía: Revista Colombiana de Geografía*, *23*(2), 45–67. https://revistas.unal.edu.co/index.php/rcg/article/view/39578

Carr, A. (2014). Is Australia a Middle Power? A Systemic Impact Approach. *Australian Journal of International Affairs*, *68*(1), 70–84. https://doi.org/10.1080/10357718.2013.840264

Chapnick, A. (1999). The Middle Power. *Canadian Foreign Policy Journal*, *7*(2), 73–82. https://doi.org/10.1080/11926422.1999.9673212

Cheng, D. (2023, October 19). *U.S. Needs to Invest More in Latin America to Counteract China in the Region*. United States Institute of Peace. https://www.usip.org/publications/2023/10/us-needs-invest-more-latin-america-counteract-china-region

China MOFA. (2016). *Basic Information about China-CELAC Forum*.

Colombo, S., López, M. P., & Vera, N. (2021). Tecnologías emergentes, poderes en competencia y regiones en disputa: América latina y el 5G en la contienda tecnológica entre China y Estados Unidos. *Estudos Internacionais: revista de relações internacionais da PUC Minas*, *9*(1), Article 1. http://hdl.handle.net/11449/206412

Comisión Económica para América Latina y el Caribe. (2015). *Informe analiza vínculos comerciales y de inversión entre Estados Unidos y la región* [Text]. Comisión Económica para

América Latina y el Caribe. https://www.cepal.org/fr/noticias/informe-analiza-vinculos-comerciales-y-de-inversion-entre-estados-unidos-y-la-region

Comisión Económica para América Latina y el Caribe. (2017). *La Franja y la Ruta es una propuesta civilizatoria de conectividad y prosperidad compartida: CEPAL* [Text]. Comisión Económica para América Latina y el Caribe. https://www.cepal.org/es/comunicados/la-franja-la-ruta-es-propuesta-civilizatoria-conectividad-prosperidad-compartida-cepal

Comisión Económica para América Latina y el Caribe. (2021). *La Inversión Extranjera Directa en América Latina y el Caribe 2021*. Comisión Económica para América Latina y el Caribe. https://www.cepal.org/es/publicaciones/47147-la-inversion-extranjera-directa-america-latina-caribe-2021

Cota, I. (2024, March 15). *The Trade Route from China to Mexico Soars by 60% in January and Establishes Itself as One of the Largest in the World*. EL PAÍS English. https://english.elpais.com/economy-and-business/2024-03-15/the-trade-route-from-china-to-mexico-soars-by-60-in-january-and-establishes-itself-as-one-of-the-largest-in-the-world.html

Cotovio, V., & Goodwin, A. (2024, February 9). *Venezuela Builds Forces Near Border with Guyana Despite Agreement to De-escalate*. CNN. https://www.cnn.com/2024/02/09/americas/venezuela-guyana-border-troops-intl-latam/index.html

Creutzfeldt, B. (2016). China and the U.S. in Latin America. *Revista Científica General José María Córdova*, *14*(17), 5–6. https://www.redalyc.org/pdf/4762/476255357003.pdf

Fitch, J. S. (1993). The Decline of US Military Influence in Latin America. *Journal of Interamerican Studies and World Affairs*, *35*(2), 1–50. https://doi.org/10.2307/165943

Foulon, M. (2015). Neoclassical Realism: Challengers and Bridging Identities. *International Studies Review*, *17*(4), 635–661. https://www.jstor.org/stable/24758570

Friedberg, A. L. (2005). The Future of U.S.-China Relations: Is Conflict Inevitable? *International Security*, *30*(2), 7–45. https://doi.org/10.1162/016228805775124589

Gamso, J., & Moffett, M. H. (2023). Unraveling the Belt and Road Initiative: China's "Building Out" Strategy. *East Asia*, *40*(1), 21–36. https://doi.org/10.1007/s12140-022-09394-1

Haglund, D. G., & Onea, T. (2008). Sympathy for the Devil: Myths of Neoclassical Realism in Canadian Foreign Policy. *Canadian Foreign Policy Journal*, *14*(2), 53–66. https://doi.org/10.1080/11926422.2008.9673463

Hakim, P. (2006). Is Washington Losing Latin America? *Foreign Affairs*, *85*(1), 39–53. https://doi.org/10.2307/20031841

Halperín, M. (2021). ALBA y Grupo de Puebla: La "verdadera" integración latinoamericana o una repetida fantasía colectiva. *Aportes para la Integración Latinoamericana, año*, *27*(45). https://doi.org/10.24215/24689912e037

Holbraad, C. (1971). The Role of Middle Powers. *Cooperation and Conflict*, *6*(1), 77–90. https://doi.org/10.1177/001083677100600108

Hua To, J. J. (2014). *Qiaowu: Extra-Territorial Policies for the Overseas Chinese*. Brill. https://brill.com/display/title/21636

Jackson, R. H. (1987). Quasi-states, Dual regimes, and Neoclassical Theory: International Jurisprudence and the Third World. *International Organization*, *41*(4), 519–549. https://doi.org/10.1017/S0020818300027594

Jordaan, E. (2003). The Concept of a Middle Power in International Relations: Distinguishing Between Emerging and Traditional Middle Powers. *Politikon*, *30*(1), 165–181. https://doi.org/10.1080/0258934032000147282

Kaska, K., Beckvard, H., & Minárik, T. (2019). *Huawei, 5G and China as a Security Threat*. Nato Cooperative Cyber Defence Centre of Excellence, CCDCOE.

King, A. (2013). Review of Review of Nationalism and Power Politics in Japan's Relations with China: A Neoclassical Realist Interpretation, by L. Y. Meng. *Contemporary Southeast Asia*, *35*(3), 459–461. https://www.jstor.org/stable/43281271

Kobek, L. P., & Caldera, E. (2016). Cyber Security and Habeas Data: The Latin American Response to Information Security and Data Protection. *Oasis*, *24*, 109–128. https://www.redalyc.org/journal/531/53163716007/html/

Lagos, M. (2023). *Informe 2023 LA RECESIÓN DEMOCRÁTICA DE AMÉRICA LATINA*. Latinobarómetro. https://www.latinobarometro.org/lat.jsp

Lee, N. T. (2020). *Navigating the U.S.-China 5G Competition*. Global China, Brookings.

Matta, J. E. (1991). Chile Y La Republica Popular China: 1970–1990. *Estudios Internacionales*, *24*(95), 347–367. https://www.jstor.org/stable/41391373

Mearsheimer, J. J. (2001). *The Tragedy of Great Power Politics*. Norton. https://catdir.loc.gov/catdir/toc/fy02/2001030915.html

Minzner, C. (2020). *End of an Era: How China's Authoritarian Revival Is Undermining Its Rise*. Oxford University Press.

Mo, J. (2016). South Korea's Middle Power Diplomacy: A Case of Growing Compatibility Between Regional and Global Roles. *International Journal: Canada's Journal of Global Policy Analysis*, *71*(4), 587–607. https://doi.org/10.1177/0020702016686380

Moravcsik, A. (2009). Europe: The Quiet Superpower. *French Politics*, *7*(3–4), 403–422. https://doi.org/10.1057/fp.2009.29

Moravcsik, A. (2010). Europe, the Second Superpower. *Current History*, *109*(725), 91–98. https://www.jstor.org/stable/45318909

Myers, M. (2018). China's Belt and Road Initiative: What Role for Latin America? *Journal of Latin American Geography*, *17*(2), 239–243. https://www.jstor.org/stable/44861544

Neto, O. A., & Malamud, A. (2015). What Determines Foreign Policy in Latin America? Systemic versus Domestic Factors in Argentina, Brazil, and Mexico, 1946–2008. *Latin American Politics and Society*, *57*(4), 1–27. https://doi.org/10.1111/j.1548-2456.2015.00286.x

North American Industry Classification System (NAICS) U.S. Census Bureau. (2022). *2022 NAICS Definition: Sector 51—Information*. https://www.census.gov/naics/?input=51&year=2022&details=51

OAS. (2003). *Cybersecurity*. https://www.oas.org/ext/en/security/prog-cyber

OAS. (2009, August 1). *OAS - Organization of American States: Democracy for Peace, Security, and Development* [Text]. https://www.oas.org/en/about/purpose.asp

Observatory of Economic Complexity. (2022). *China (CHN) Exports, Imports, and Trade Partners*. The Observatory of Economic Complexity. https://oec.world/en/profile/country/chn

Observatory of Economic Complexity. (2023). *México (MEX) Exportaciones, importaciones y socios comerciales*. Observatorio de Complejidad Económica. https://oec.world/es/profile/country/mex

OECD. (2019). *The Road to 5G Networks: Experience to Date and Future Developments*. OECD. https://doi.org/10.1787/2f880843-en

Office of the United States Trade Representative. (2023). *Countries & Regions*. United States Trade Representative. https://ustr.gov/countries-regions

Özşahin, M. C., Donelli, F., & Gasco, R. (2022). China–Turkey Relations from the Perspective of Neoclassical Realism. *Contemporary Review of the Middle East*, *9*(2), 218–239. https://doi.org/10.1177/23477989211062659

Pew Research Center. (2023). U.S. Global Image—Research and data from Pew Research Center. Pew Research Center. https://www.pewresearch.org/topic/international-affairs/global-image-of-countries/us-global-image/

Polga-Hecimovich, J. (2022, November 22). *China's evolving economic footprint in Latin America*. https://www.gisreportsonline.com/r/chinas-economic-power-grows-in-latin-america/

Press-Barnathan, G. (2006). Managing the Hegemon: Nato under Unipolarity. *Security Studies*, *15*(2), 271–309. https://doi.org/10.1080/09636410600829554

Progress at Foro De Sao Paulo. (2002). LADB. University of New Mexico. https://digitalrepository.unm.edu/cgi/viewcontent.cgi?article=10037&context=noticen

Rivera-Matias, N. (2022). *Did Costa Rica's Decision to Recognize China Pay Off?* The Diplomat, Online. https://thediplomat.com/2022/11/did-costa-ricas-decision-to-recognize-china-pay-off/

Robertson, J. (2017). Middle-Power Definitions: Confusion Reigns Supreme. *Australian Journal of International Affairs*, *71*(4), 355–370. https://doi.org/10.1080/10357718.2017.1293608

Rogers, R. I. (2022). *The Switch: Why Do Panama and Nicaragua Now Recognize China over Taiwan?,* Modern Languages, Philosophy and Classics Theses. 4. https://docs.rwu.edu/foreign_languages_theses/4

Sabatini, C. (2013). Will Latin America Miss U.S. Hegemony? *Journal of International Affairs*, *66*(2), 1–14. https://www.jstor.org/stable/24388282

Sahd, J., Zovatto, D., & Rojas, D. (2024). *Riesgo Politico America Latina.* Centro UC Estudios Internacionales. https://centroestudiosinternacionales.uc.cl/images/publicaciones/publicaciones-ceiuc/2024/Riesgo-Politico-America-Latina-2024_compressed.pdf

Sanchez-Lopez, M. D. (2023). Geopolitics of the Li-ion Battery Value Chain and the Lithium Triangle in South America. *Latin American Policy*, *14*(1), 22–45. https://doi.org/10.1111/lamp.12285

Schenoni, L. L., & Mainwaring, S. (2019). US Hegemony and Regime Change in Latin America. *Democratization*, *26*(2), 269–287. https://doi.org/10.1080/13510347.2018.1516754

Segovia, D. (2013). Latin America and the Caribbean: Between the OAS and CELAC. *Revista Europea de Estudios Latinoamericanos y Del Caribe / European Review of Latin American and Caribbean Studies*, *95*, 97–107. https://www.jstor.org/stable/23595694

Shoujun, C., Zheng, Z., Baiyi, W., Dongzhen, Y., Jingsheng, D., Jianmin, Y., Zhenxing, S., Shixue, J., Yinghua, Z., Fan, Z., Haibin, N., Xiaodai, X., Cunhai, G., & Shuangrong, H. (2018). China Y La Infraestructura En América Latina Desde La Perspectiva De La Diplomacia Económica. In W. Baiyi (Ed.), *Pensamiento social chino sobre América Latina* (pp. 261–290). CLACSO. https://doi.org/10.2307/j.ctvnp0jw3.14

Silver, L., Huang, C., & Clancy, L. (2022, September 28). *How Global Public Opinion of China Has Shifted in the Xi Era.* Pew Research Center's Global Attitudes Project. https://www.pewresearch.org/global/2022/09/28/how-global-public-opinion-of-china-has-shifted-in-the-xi-era/

Takahashi, R. (2023). *Analysis of the Theory That Latin America Is Turning into "China's Backyard" and the Outlook for the Future*. Mitsui & Co. Global Strategic Studies Institute. https://www.mitsui.com/mgssi/en/report/detail/__icsFiles/afieldfile/2023/09/19/2308k_takahashi_e.pdf

Tekir, G. (2020). Huawei, 5G Network and Digital Geopolitics. *International Journal of Politic and Security*, *2*, 113–135.

The Belt and Road Initiative in Latin America: How China Makes Friends and What This Means for the Region. (2022, March 18). Latin American Focus Group. https://blogs.eui.eu/latin-american-working-group/the-belt-and-road-initiative-in-latin-america-how-china-makes-friends-and-what-this-means-for-the-region/

Tiezzi, S. (2020). *China's Bid to Write the Global Rules on Data Security*. The Diplomat. https://thediplomat.com/2020/09/chinas-bid-to-write-the-global-rules-on-data-security/

Tomasic, R. (2015). Company Law Implementation in the PRC: The Rule of Law in the Shadow of the State. *Journal of Corporate Law Studies*, *15*(2), 285–309. https://doi.org/10.1080/14735970.2015.1044769

Trubowitz, P., & Burgoon, B. (2022). The Retreat of the West. *Perspectives on Politics, 20*(1), 102–122. https://doi.org/10.1017/S1537592720001218

Umbach, F. (2020). *Eu Policies on Huawei and 5g Wireless Networks: Economic–Technological Opportunities Vs Cybersecurity Risks.* S. Rajaratnam School of International Studies. https://www.jstor.org/stable/resrep28286

Urdinez, F., Mouron, F., Schenoni, L. L., & de Oliveira, A. J. (2016). Chinese Economic Statecraft and U.S. Hegemony in Latin America: An Empirical Analysis, 2003–2014. *Latin American Politics and Society, 58*(4), 3–30. https://www.jstor.org/stable/44683858

Urdinez, F., & Winters, M. (2021). *COVID-19 Diplomacy and Soft Power: Did Vaccine and Equipment Distribution Improve Perceptions of China in Latin America?* https://www.furdinez.com/uploads/2/2/5/6/22565746/working_paper_-_urdinez___winters_2021_oct.pdf.

U.S. Bureau of Economic Analysis (BEA). (2023). *U.S. Direct Investment Abroad: Balance of Payments and Direct Investment Position Data.* https://www.bea.gov/international/di1usdbal

Vallejo, M. C., Espinosa, B., Venes, F., López, V., & Anda, S. (2019). Evading Sustainable Development Standards: Case Studies on Hydroelectric Projects in Ecuador. In *Development Banks and Sustainability in the Andean Amazon.* Routledge.

Venturini, J., & Garay, V. (2021). *Reconocimiento facial en América Latina: Tendencias en la implementación de una tecnología perversa.* Al Sur. https://www.alsur.lat/sites/default/files/2021-11/ALSUR_Reconocimiento_facial_en_Latam_ES.pdf

Vidal, F. (2023). The Antarctic Peninsula: Argentina and Chile in the Era of Global Change. *The Polar Journal, 13*(1), 13–30. https://doi.org/10.1080/2154896X.2023.2205236

Wang, C. N. (2023, December). *Countries of the Belt and Road Initiative (BRI).* Green Finance & Development Center. https://greenfdc.org/countries-of-the-belt-and-road-initiative-bri/

Woo, F. J. (2011). Review of Review of On China, by H. Kissinger. *China Review International, 18*(2), 200–204. https://www.jstor.org/stable/23733433

Wu, M. (2016). The “China, Inc.” Challenge to Global Trade Governance. *Harvard International Law Journal, 57.*, 261–324. https://www.law.berkeley.edu/wp-content/uploads/2020/05/WuMark.pdf

Xinsheng, Z. (2006). El entendimiento sobre el Tratado de Libre Comercio entre China y Chile. *Estudios Internacionales, 38*(152), 113–121. https://www.jstor.org/stable/41391860

Zaballos, A. G., Rodriguez, E. I., Kim, K. W., & Park, S. (2020). 5G: The Driver for the Next-Generation Digital Society in Latin America and the Caribbean. *IDB Publications.* https://doi.org/10.18235/0002264

4 International Relations Theory and the Dissemination of Technology

The massification of specific technologies that involve high traffic of personal information—like 4G and 5G networks—around the world has raised various questions in international relations studies. Why would or should a state be interested in having domestically produced technologies spread worldwide? Coincidentally, why should receiving countries be concerned about the country of origin of an innovation they decide to acquire and disseminate around the country?

Traditionally, technology has been understood as an asset in building up to achieve dominance and power. The condition by which certain technological advancements are deemed valuable to a nation's military prowess remains contingent upon their practical utility. During the First World War, an arms race between that era's leading powers was supported by rapid technological progress, and few reject the connection between the increase in armament and the tensions that led to the conflict.[1] The Second World War also witnessed technology-driven outcomes, and the creation of the atomic bomb, followed by the decisive impact of nuclear proliferation and intercontinental ballistic missiles, had an equivalent effect on the way international relations were analyzed during the 20th century.[2] Game theory—in several models—[3] has been critical in analyzing how actors behave with limited knowledge and portraying the fragile nature of power balance, particularly with nuclear deterrence.[4] Deterrence is not determined by technological capabilities alone but also by factors such as the adversary's credibility, uncertainty, and balance achieved through politics.

According to a realist perspective, actors act and react in an anarchic world where technology gives them relative power to persuade or coerce states. Nevertheless, since the rise of globalization and the incidence of technologies that are not directly linked to specific states, the basic assumptions of some traditional analyses have become blurred. In addition, the internet's interconnectivity has long been the source of international debate, not only related to governance and accountability but also to the shared rules that states must agree to, as well as international standards.[5] An example of such standards is the Budapest Convention on Cybercrime, which seeks to facilitate cooperation among states when needed. The convention has been signed by 69 parties, with an extra 22 being invited to join. Its signatories include the US, Japan, Israel, the European Union, and more than eight countries from Latin America,[6] which sets a basis for deducing what the political and

DOI: 10.4324/9781003489450-5

philosophical ideas of this set of countries are compared to those not in this list. In other words, some could analyze these international agreements and establish a line between the idea of a global and shared internet and the consistent division between blocs that support particular views on what the internet should be. The case of international technology transfers is similar. The decision on which provider to choose depends not only on price and technical quality but also on political implications. Whether these processes result from a multipolar spread of power or several powers aligning themselves behind security or value-related principles widely depends on the target country and the case study.

Despite all those facts, the academic literature on science and technology research in international relations is limited compared to other areas[7] or rather spread in focus. Coincidentally, all main theories of international relations seem to have limitations when analyzing hybrid affairs. In this chapter, I will revisit and explore the concept of power and its relationship with technology—including the incorporation of soft power. Drawing from other disciplines, I explain how some states may benefit from the spread of technology and discuss the inherent limitations of international relations in understanding these processes. I also identify the framework and assumptions this book is based on. By doing this, readers will receive a reasonably clear picture of how to interpret technology transfer in terms of theoretical and evidence-based points of reference.

International Relations, Power, and Technology

In Chapter 2, I mentioned that "power" is crucial to understanding why technology is so essential for countries to invest and seek influence. Although this definition can vary depending on the analysis framework, there are some clear lines in the discussion of technology transfers. First, it is essential to understand that technology is, in many forms, a central element in the calculations of power balance in international relations. These material considerations impact state decisions and the outcomes of power competition from a traditional perspective. It is then agreed that when technology is not used directly in confrontations, it is essential for states to escape the physically coercive actions—that Morgenthau had already distanced from power.[8] Some authors—such as Gilpin—directly refer to technology as a form of control, and its expansion can be part of a state's capacity to change the international order or enlarge its influence.[9] While technology can be connected to material power in the form of military innovation, countries might also identify value in the diffusion of innovation in terms of diplomacy and political influence.[10] In this sense, transferring technology to a foreign country impacts a country's wielding of influence abroad.

In measuring power in different indexes, the capacity to produce and wield technology often appears as a core element or dimension—but not always. For example, there is a debate on whether the Composite Index of National Capability (CINC), which mainly measures hard power and industrial output, is a proper standard for comparing countries' capabilities in a post-soviet-collapse scenario.[11] On the other hand, the Global Power Index (GPI) directly incorporates technology into the power calculation, considering the investment in research and development.[12]

Considering this, a key point of debate is the capacity to compare technological power between states based on a country's actual innovation or investment in research and development. The reality is that developing and *poor countries* can take advantage of the advancements, investments, and overall experience of more powerful states[13] that have already sacrificed resources to acquire said technology. Both institutions and strategies can be replicated and used as the base for developing other states that will not need to deploy the same time and resources to access innovation—something technology transfers can be especially relevant for. Even if countries are not at the forefront of research and development, there are tools for them to either replicate or adopt new technologies. Gerschenkron coined this as "backwardness advantage" in his 1962 work: "Economic Backwardness in Historical Perspective." In his view, industrialized countries in later stages had immediate access to tools, technological progress, and policies that could simplify the process and reduce the costs and risks of investing in large-scale technological transformation.[14] In other words, undeveloped countries could import technology to compensate for their backwardness, and this can lead to an advantage because their production costs would become cheaper when competing with more developed countries. A clear example of this could be the Japanese and Korean cases during the late 20th century and, more recently, the case of China's rapid innovation wave. However, Asian success stories also showcased a series of parallel policies in agriculture and productivity that cannot be ignored in the broader analysis.[15]

In simpler terms, countries that are not traditional innovators can rapidly catch up and, in some cases, utilize technology transfers to gain material power for themselves—or even to compete with existing technology providers. In this context, other indicators, such as reputation, political power, and influence, might directly involve wielding power through innovation. Joseph Nye defines power in three different forms. First is hard power, linked with the capacity to coerce with force or economic incentives. Then soft power—the most relevant to this book—is the ability to affect an outcome through reputation and attraction rather than threats; it "rests on its resources of culture, values, and policies."[16] The third one, smart power, "combines hard and soft power resources. Public diplomacy has a long history of promoting a country's soft power and was essential in winning the Cold War."[17] Drawing from the interaction of domestic and international factors, as well as the need a state has to assert its position in the global theatre, soft power and sharp power are highly relevant to the debate of technological transfers and to answer the initial questions on incentives states might consider disseminating their domestically produced intellectual property. At the same time, the interaction of private entities makes perfect sense in a scenario in which they are part of a country's reputation-building scheme, mostly when the government has a direct influence and power in promoting this strategy. When the product of innovation is a multilevel endeavor that includes several governments, different companies, and members of civil society across borders, this definition becomes more difficult. Still, it remains valid if the incumbent actors share political and ideological elements that make the coordination of the dissemination process one that improves all parts' soft power.

The dissemination of COVID-19 vaccines during the pandemic—an affair that required highly intense negotiations among nations—is a clear example. Around the world, there were technological, scientific, and political underlying discussions in the standard the vaccines followed[18]. For example—and though this is only indirectly related to technology transfers in terms of infrastructure and knowledge—China promoted its vaccine, SINOVAC/CANSINO, across less developed countries that could not reach the supply of traditional Western vaccines created by private endeavors. Russia did the same, reaching economically constrained countries at the beginning of the outbreak with its *Sputnik* alternative. This decision aimed to develop an international perception of scientific superiority based on state support. In the case of China, the People's Liberation Army was involved in creating and distributing the vaccine.[19] In the meantime, Western nations advocated for joint solutions that heavily involved the private sector and international efforts. Pfizer/BioNTech was a German American vaccine, AstraZeneca/Oxford was British, Swedish, and American, and Moderna was a private alternative developed in the US with public support. Their *modus* of distribution, in the form of COVAX, was intended to ship these vaccines worldwide.

Even though these Western alternatives were proven more effective than the Russian[20] or Chinese alternatives, the discussion of which vaccine to choose was political. Latin American nations became ideal candidates for technological competition, as they are culturally aligned with the West while lacking resources and a vaccine alternative. Argentina, a long-standing democracy, received Russian vaccines as an affordable option.[21] Chile acquired millions of Chinese vaccine doses, then donated to other countries in the region, including Paraguay, a country that still recognizes Taiwan.[22] Interestingly, while doing this, Chile re-vaccinated its population with Pfizer and other established Western alternatives. Consequently, the vaccination process was dominated by actors that targeted the population of Latin American countries, competing outside of confronting material-power dynamics.

In this case, domestic conditions, economic incentives, and ideological debates were intertwined with the most basic need for a country's survival, and this particular example deals with the non-traditional nature of private companies. In addition, the challenging position of actors such as the US in pursuing a mechanism of vaccine donation contrasts with the Russian and Chinese alternatives through COVAX.[23] Both the US and their allies, as well as China, had normative and political incentives to send vaccines to less-developed countries. That form of international policy, which is no war nor peace, becomes even more complex with the transfer of knowledge and technology.

If we are heading—although this concept is still a matter of debate—into a new Cold War, and following the existing knowledge on this matter, then expressions of soft power are playing a central role in state interaction among the diffuse borders and digital behavior. Also, in the quest to enlarge said influence, countries adapt their strategies. The nature of the states sometimes gives the capacity and flexibility of said adaptation, as well as the context. Companies act on self-interest, but states can enact policies that make specific markets more attractive than others and create instances that help promote their industries abroad. On the contrary, private companies are subjected to different treatments depending on what country they are most related to. The case of Australia's ban on Chinese companies in telecommunications processes is relevant to this book and context. In 2024, Chile excluded Israeli companies[24] from its International Air and Space Fair (FIDAE in Spanish), the oldest Latin American conference held once every two years to showcase the latest innovations in the sector, including civil and military aviation, as well as security and cybersecurity in the sector.[25] This is directly related to Chile's support for Gaza in the conflict that began with the Hamas attack in October 2023, and that has since been highly controversial internationally. Chile, a country with a 500-thousand-people Palestinian community, has reasons that go beyond security and logic to make this particular decision against private companies. In many aspects, Gaza has built strong sympathy across the Chilean population. Therefore, this controversial decision had supporters who were also in power at the moment, which affected Israeli companies' prospects of transferring technology.

Soft Power and Values

The difference between a state showing its values and their effect on soft power is a country's active attempt to portray a reliable image, which includes ethical approaches but goes further. In short, it is more related to the image a country makes to the outside world than what its people do or think.

Understanding and utilizing soft power is essential in shaping a country's interactions with the world. According to McClory and Harvey, there are three reasons for this. First, soft power provides flexibility in a constantly changing international landscape, particularly during the ongoing digital revolution. Second, it is objectively less expensive to nurture than hard power. Finally, negotiating and expecting a positive collaborative response is more convenient and effective in a multilateral scenario.[26] States should use their existing capabilities and resources to shift others' values toward their priorities. The Soft Power 30 index, which measures a country's ability to attract and persuade, is a valuable framework to assess the role of technology in constructing soft power. Being able to convince other states is a valuable capability when shifting norms, as is the case with cybersecurity standards and freedom of the internet.[27] Jhee and Lee separate the measurement of soft power into two dimensions: normative through legitimacy and affective through attraction. The first relates to non-aggressive political influence, such as international assistance and observation of international norms. In contrast, the second refers to culture, development, education, and scientific advancement.[28] In this line of thought, technology touches directly and indirectly in both dimensions, and these will be considered in later chapters as I review cases in which both expected and unexpected outcomes were achieved.

In the case of the transfer of technology, the incorporation of soft power helps explain the outcomes of processes in Latin America that might initially seem counterintuitive. In particular, the prominent role of private companies that garner a reputation that is somehow associated with their country of origin—thus influencing how decision-makers and the public perceive that state.

Traditional Interpretations of Technology

There is an ongoing discussion on the inclusion of new technologies across theoretical frameworks in international relations. While authors such as Weiss have debated the relationship between notions of power and technology,[29] Eriksson and Giacomello proposed a thorough overview of possible interpretations of the main theories in IRs,[30] something recently re-iterated in Eriksson and Newlove-Eriksson.[31]

Realism, a theory that considers international politics as an occurrence based on the laws of nature,[32] opposes idealism in analyzing international relations.[33] In other words, states will always seek their survival, and this is an end to the power-seeking interactions with different units. Realism is often considered a more scientific approach to studying international relations since it intends to describe scenarios in their analysis without a moral view or direction. At the same time, the description of outcomes is considered more fatalistic or pessimistic when states interact.[34] Realism can be helpful for researchers who study the potential threats of introducing a new technology to a system. In interstate conflicts, the direct relationship between technology and the accumulation of power provides the basis to sustain this connection. The assumption of an anarchic world in which states compete and are constrained by a set of conditions—under Waltz's approach—[35] puts a state in front of a set of decisions that affect their role in the balance of power in this international sphere. Critics of *orthodox realists*[36] point to the fact that some existing occurrences fall out of the scope of this theory, and they are dismissed rather than approached by researchers, creating gaps and contradictory positions, such as an explanation for the continuity of NATO after the fall of the Soviet Union.[37]

Therefore, this theory can explain technology in power competition. Still, when theorists are inflexible in their analysis or generalization, they struggle to incorporate external elements that are not directly related to traditional power[38] or take agency from states as the sole units of study. This critique also transpires in the negotiations and dissemination of technologies less related to traditional material power. Since states are units, domestic or secondary affairs are not considered in the outcomes of a particular situation, nor are decisions made for value-related or ideological motivations.[39] Despite the increasing openness of realists to consider alternative factors, the scientific approach to analysis, while useful in many scenarios, may prove inadequate in specific contexts. In other words, when analyzing a state's decision-making concerning an issue related to international relations, the realist theory will tend to isolate factors to approach the problem, which takes away attention from relevant elements that help explain the outcome.

But perhaps the most relevant critique of realism as a perspective to analyze technological transfers is that realists are said to be "remarkably 'unworried' about contemporary technological trends,"[40] except when it comes to Weapons of Mass Destruction (WMD).[41] Again, realists like Morgenthau or Herz recognize the role of technology to some extent, as long as it has a decisive role in direct conflict.

Constructivism critiques realist approaches to international relations and responds to "neo" conceptions of liberalism and realist theories.[42] Its contribution is significant because it uniquely captures contextual elements that explain a country's behavior rooted in social norms and relationships.[43] This has been particularly pertinent in the post-Cold War era, where globalization has encouraged cooperation and trade over military disputes and the emergence of international institutions that further promote a stable international order enabled by connectivity, business, and technology.[44]

In analyzing technology and IRs, this theory emphasizes the nature of social relations. A well-known comparison that Alexander Wendt makes to explain the issue is that "500 British nuclear weapons are less threatening to the US than five North Korean Nuclear weapons."[45] In other words, constructivists claim to analyze the contexts in which two actors are intertwined and, when addressing technology, the capacity of a state is not immediately disregarded as a threat but the intentions and norms that overlook that relationship.

Since constructivism has a normative intent, the theory is also commonly used to explain the incentives countries might have to approach and comply with international regulations and directly change certain state behaviors.[46]

Nonetheless, there are several limitations to the constructivist approach to this phenomenon. The theory is criticized for dismissing the empirical problem of uncertainty, which is critical in understanding state decision-making.[47] First, when it comes to technology transfers across countries, uncertainty is always present since countries are technically relying on an external actor to provide a nationwide service or infrastructure network. Secondly, some states—including most in Latin America—face a lot of limitations that are not related to their values, social relations, and the norms they aim to adhere to. Incentives are sometimes external or provided by the restraints of the very nature of a state's capabilities. In other words, a provider might be chosen merely because of the state's economic limitations, not by the shared values, its reputation, or its position on an international scale.

Similarly, actors will not only act and react based on the needs of their social environment but primarily on their interests. The novelty of digitalization and current technologies is that those interests are unclear and involve many other factors. A state can negotiate the introduction of an optic fiber cable with the incentive to benefit businesses and increase connectivity in rural areas that would otherwise be less approachable to the state. When deciding which provider a state will consider implementing a project, there will be economic incentives—as in any other infrastructure project—and also the extent to which the state can regulate cybersecurity standards. All those factors are related to domestic decisions but intertwined with a foreign actor. In the case of norms, incentives also influence whether a country will adopt international standards.

Moreover, incentives are sometimes the result of established relations but economic incentives or the lack of resources. That is the case with several technologies and scientific projects introduced in Latin America. Countries that seemingly have all incentives to cooperate to maintain alliances do not have the capacity or will to do so. In contrast, others seized the gap and weaponized the distribution system for convenience.

Liberalism is the third major theory used to explain phenomena in international relations research. As it reflects the values of classical liberalism, this theory assigns a philosophical and deontological position to what states should do in the international system. For example, republican liberalism, or the idea that democracies tend to be more peaceful than other forms of government—as the democratic peace theory develops; commercial liberalism, as the idea that interdependence fosters cooperation; and regulatory liberalism, as the conceptualization of how institutions and law can shape the international order.[48]

Both liberalism and constructivism are regarded as "utopian" by realists for their moral pursuit of shaping the international sphere,[49] but liberal writers are often proclaimed anti-utopians.[50] Coincidently, the liberal theory is said to accept some fundamental ideas from realism—including the notion of egoism—even though these two postulates are often theorized as opposites by realists.[51] Keohane and Martin argue that realism and institutionalism (linked with liberalism) are rationalistic at their core and serve a utilitarian purpose.[52] They also propose that the difference between both theories is the attitude or scope of understanding the international sphere. Realists are more pessimistic and grimmer about global politics, while institutionalists look for solutions to the issues it encompasses, recognizing a state's interest and the value of cooperation.[53]

Liberalism is also keen on putting the individual and their rights at the center of international relations dynamics. Since peace and freedom are regarded as conditions for human prosperity, there is a normative objective in the interactions of nations.[54]

The existence of an "international order" and, related to the matter of this book, an "internet global standard" are based on the idea of cooperation among nations as a means to the end of a "somewhat" better outcome but are constantly being challenged by emergent powers that also question the fundamental values of the said liberal order, such as human rights and sovereignty over the internet.[55]

Those normative and deontological aspects of this theory are also the main critiques of its relevance for understanding the phenomenon of technology transfers. First, economic interdependence is currently being challenged by the increasingly securitized international environment, first regarding the use of commercial constraints to shape security,[56] and second, since there are records of the use of technological and economic tools to take advantage of the system.[57] This book, in particular, lies on the assumption that factors outside of cooperation and economic interdependence guide the way Latin American states behave when a new technology is introduced. Second, institutions are being co-opted by countries that challenge the international order and have their normative agenda. "Cyberspace"—sometimes referred to as the fifth domain of warfare[58]—is one of those areas since there have been various attempts at coordinating internationally toward a shared internet that is open to access, but at the same time, raising concerns across liberal democracies that initiatives such as providing the International Telecommunications Union (ITU) with regulatory capacity would enable more authoritarian nations to align toward a more censored and less tolerant online governance.[59] In other words, cooperation and collaboration imply giving in to the requests of ideas that challenge the normative and more optimistic position of liberalism. This is also the case in the candid dispute of 5G networks, where two blocs are being defined not only by the quality and price of the products the companies manufacture but also by the political position of the actors that have a direct influence over those companies or the countries that are to implement them in their domestic environments. While liberalism can explain and link IRs research areas not reached by realists,[60] it can also struggle when drawing a line between cooperation and economic interconnectedness and the effects of technology in a domestic and international environment.

The Theoretical Limitations and Drawing from Other Disciplines

International relations studies have explored the relationship between nation-states and the dissemination of innovation through various theoretical frameworks, such as realist, liberal, or constructivist. These frameworks draw parallels between traditional power structures and the impact of technological innovation on international dynamics. Examples of such effects include the introduction of warfare technology, which changed the outcomes of the world wars in the 20th century, as well as technological advances in nuclear science, the internet, artificial intelligence, espionage, surveillance, cyber threats, data transfer, and innovation transfer since the 1980s.[61]

Leaning on the initial question of why transferring technology is essential, theories have different perspectives on the priorities a state should have regarding international relations. Part of the difficulty layer that technology adds is the understanding of new threats since the concept of "hybrid warfare" is growing in traction to evaluate how to respond to potential attacks that don't have a physical component but are digital or sociological. Disinformation is one such form of active engagement that aims to achieve a more aggressive objective and use hackers to intercede abroad.[62] In addition to that, a country might have long-term goals in the dissemination of technology originated or produced in their territory, which might range from economic benefits derived from standard imposition and harmonization[63]—or adapting to external norms to get benefits such as lower costs and higher-level technology—to the strengthening of their public diplomacy and reputation within the population of the target state. In addition public opinion can intertwine with a state's agency toward a nation-scale decision on technology acquisition.

Because of all these elements and the complexity of these problems, a neoclassical realist framework, supported by the ideas of soft power and two-level games, can better examine the issue of technology transfers to Latin America. The cases I analyze in later chapters indicate that while necessary, security does not play a role in every decision involving China and the US or two eventual blocs of a new Cold War. However, I present notable examples of direct opposition between both powers In one of the cases, Argentina is said to have given away part of its sovereignty to Beijing in a deal that has widely been considered asymmetric, non-transparent, and concerning to Argentina's security.[64] Domestic security considerations were discussed in other scenarios, such as the public tender for the passport system that put Chile in the middle of a power struggle between the US and China. Moreover, the overarching tensions between the nation's two leading economic powers dominated the public debate.

Neoclassical Realism as an Answer to the Limitations

In this book, I propose that the neoclassical realism approach addresses complex interacting factors, especially in medium and small states that tend to follow regional and international trends.[65] While understanding the role of security and the idea of an anarchic international theater, it also recognizes the role of domestic constraints and elite perceptions in shaping foreign policy decisions.[66] Neoclassical realism also integrates systemic and unit-level variables in a deductively consistent manner. It can explain how broad international trends and specific domestic factors influence foreign policy decisions.[67] It recognizes domestic constraints and the role of elite perceptions in shaping these decisions, withdrawing from Waltz's theory. For some scholars, this means that they will not abide by the principle of a rational state with no internal constraints,[68] which is the main critique this theory gets from more classical realists.

According to Ripsman, this approach is convenient for understanding the subtleties of a given context because it allows for revisions and considers other fundamental principles from liberalism and constructivism. Ideas like identity, values, and economic incentives can indirectly influence a country's foreign policy because they may align with the interests of domestic actors.[69] On that same line, Coetzee and Hudson note, in the context of democratic peace studies, that:

> Through an incorporation of variables stretching across differing levels of analysis, neoclassical realism not only highlights the complexity of theorizing foreign policy behavior but also points to the complex causal interaction of systemic and unit-level variables.[70]

Gourevitch explains that the relationship between domestic and system-level variables must be critical. Neoclassical realism scholars accept that the international economic system or international institutions do not affect policymaking in a country.[71] A state might adapt its legislation to match the requirements of an international convention, even if it is not compulsive or binding. This can quickly occur in technology when there are indirect benefits from adopting the legislation, such as a more accessible entry of products and services to a foreign market or a more economically sound investment in the domestic space. Coincidentally, state practices can be used as a basis for the accorded international standards. In the case of network neutrality or non-discrimination, the idea of non-interference is vital in the US's private environment. Still, it has been replicated in several countries around the globe and is particularly relevant in Latin America. Non-discrimination is a matter of candid international debate since it depends on participation in the internet market and the level of access providers have to digital traffic.[72]

Considering that, in his work, "The Second Image Reversed," Gourevitch is emphatic in debunking the top-down relation between the international system and a country's domestic policy or between strong and weak countries.[73] His conclusions note that these case studies should be interpreted to understand the thick interdependence of international and national factors and analyze cases as "a whole" rather than trying to separate their factors.[74]

Attributing a country's decision to a single variable is an oversimplification of a complex phenomenon because it is necessary to consider the system and context. Therefore, it would be erroneous to hastily conclude that the US or China dominates technology based solely on their economic or political interdependence, especially in telecommunications.

Technology adoption within a country is often conducted through institutional processes or negotiations. I propose that Robert Putnam's "two-level games" theory frames how domestic and international factors shape international debates in the context of technology and go further than classical state-to-state negotiations.[75] Briefly recalling his proposal, Putnam establishes that these negotiations are conducted on two levels: the international level, where countries negotiate, and the domestic level, where politicians and other domestic actors attempt to influence the negotiations. He argues that, for international negotiations to be successful, both levels of the game must be considered. At the international level, countries must reach a mutually beneficial agreement that satisfies all parties' interests. At the domestic level, politicians and other domestic actors must be able to persuade their constituents to support the agreement and convince other domestic actors, such as interest groups and the media, to back it as well.[76] It is essential to add that some authors dismiss Putman's "two-level games" as a pure neoclassical realist proposal but rather one more closely related to liberalism.[77] It is true that since neoclassical realism accepts certain conditions by other theories, it is, in some cases, difficult to draw a clear line between theoretical frameworks.

Nevertheless, these criticisms mirror those given to the neoclassical theory in general, and the author is recognized, along with Ripsman and Rose, as neoclassical realism theorizers, despite the ongoing debate on the lines of priorities this theory should apply to the factors involved in the analysis.[78] At the end of the day, national actors will be the ones who directly interact with said technologies at an institutional or personal level. A state administration understands the international context and the interests, legal status, rights, values, and economic capabilities of the final users.

Civil Society in Neoclassical Realism

Considering the importance of domestic actors in developing this scope for research, I find it necessary to point out that civil society in Latin America plays a vital role in decision-making[79] and policymaking. Civil society has a lot of definitions that end up being instrumental for the ideology that frames the concept within a puzzle.[80] In the case of South American countries, such as Chile, these organizations have a recognized role in issues such as democratization and economic liberalization, as they take up roles traditionally managed by the state.[81] Intermediate groups that are separated from the political sphere have a say in the pathways of the domestic politics of their countries and, therefore, will be affecting foreign policy indirectly and directly. A chamber of commerce, for example, might coordinate and influence the state in their foreign policy, either toward a more liberal or protectionist approach, depending on the needs of those civil society institutions. Similarly, a state might be compelled to take sides or speak out on a particular international issue if it is related to or active in the domestic realm.

The International Development Bank, which is to this date related to infrastructure, development, and modernization in Latin America (and therefore relevant to this research), has long promoted the role of civil society in democracies.[82] Larry Diamond has long addressed the importance of these groups in democratic consolidation and governance. He also expands the definition to other actors, denying that "civil society" is just a synonym for non-profit organizations or society itself:

> Civil society is an intermediary entity, standing between the private sphere and the state. Actors in civil society need the protection of an institutionalized legal order to guard their autonomy and freedom of action. Civil society encompasses a vast array of organizations, formal and informal. These include groups that are: 1) economic; 2) cultural; 3) informational and educational; 4) interest-based; 5) developmental; 6) issue-oriented; and 7) civic. In addition, civil society encompasses "the ideological marketplace" and the flow of information and ideas. This includes not only independent mass media but also institutions belonging to the broader field of autonomous cultural and intellectual activity.[83]

These arguments are coherent with the idea of an assortment of factors impacting decision-making in South America and exhibit the need to go further than an economic or security and a state-to-state relationship. In the case of this research. civil society's role is instrumental in understanding why these countries have chosen a standard for specific technologies. In that same spirit, countries with more restrained or less developed civil participation could be expected to behave differently than those with democratic and active citizen participation.

Technology Research in Neoclassical Realism

I previously mentioned that technological adoption often involves private actors in the form of individual innovators, companies, or stakeholders—to which neoclassical realists provide an answer. Still, insufficient research has been conducted to analyze technology and its standardization in formal studies. Following existing theoretical discussions, technology dissolves the net power of a state.[84] When authors attempt to oversimplify their analysis for a more dualistic analysis, often in a more traditional theory, the complexity of technological development as a phenomenon is lost.[85] Taliaferro established that technologies worldwide—due to anarchic competition—will end up "being similar" across states. In other words, manufacturers will try to compete with and emulate successful innovation outputs, incentivizing nations to adapt to international innovation trends for military purposes and domestic development.

Telecommunications is a clear and crucial example of this, as it can be related to technology and public policy. On the one hand, the proper design of critical policy allows foreign actors to provide the technology. Still, the opinions of current stakeholders in the domestic sphere need to be addressed, and some of these will be technological companies and politicians. On the other hand, the state will assess the issue from a cost and benefits position, where factors such as diplomatic relations or security provisions are put into the calculations. Again, Putnam's input on this issue is valuable, as his proposal acknowledges the interrelation of factors and assesses its complexity in the context of negotiations and decision-making.[86]

There is a final point to which I find a strong connection between neoclassical realists and nationwide technology implementation. While security is often measured against deterrence or offensive capacity, a growing concern is related to the vulnerability and weaponization of technologies unrelated to defense capabilities. On the one hand, cybersecurity creates a gray zone of what can be considered aggression at an international scale, with countries increasingly calling for a coherent definition and response, such as establishing new domains of security,[87] cognitive warfare,[88] or hybrid warfare. Technologies that are destined for domestic and civil use can also be repurposed for other endangering objectives. For example, Australia has prevented Chinese companies from participating in public tenders related to telecommunications because of security concerns, which kickstarted a political-legal debate related to WTO compliance and strong criticisms from the Chinese state.[89]

Notes

1 Houweling and Siccama (1981), p. 157.
2 Skolnikoff (1994), p. 224.
3 Game theory is general, but analysts normally analyze several strategies that the actors use to face uncertainty. In the case of Powell, said model is Brinkmanship, which gives states the ability to decide how long it is willing to hang on to the strategy (Powell, 2003, p. 94). In Chapter 2, I also mentioned the "prisoner's dilemma" and "The Richardson model," which provide other conditions in which actors take decisions.
4 Powell (2003), pp. 88–89.
5 Haass (2020), pp. 202–203.
6 *Budapest Convention—Cybercrime* (n.d.).
7 Weiss (2005), online.
8 Morgenthau (1948), p. 13.
9 Gilpin (1981b).
10 Morgenthau (1955), p. 129, as seen in Berenskoetter and Williams (2007), p. 49.
11 *Composite Index of National Capability by Country 2024* (2024); Heim and Miller (2020), p. 6.
12 Heim and Miller (2020), p. 7.
13 Ferguson (2012), p. 424.
14 Gerschenkron (1962), pp. 6–10.
15 Gunnarsson (2016).
16 Nye (2008), p. 94.
17 Nye (2008), p. 94.
18 Urdinez and Winters (2021).
19 Kania and McCaslin (2020), p. 24; Zhao et al. (2020).
20 De Soto (2021), pp. 411–412.
21 Malamud and Núñez (2021).
22 Malacalza and Fagaburu (2022).
23 Malacalza and Fagaburu (2022).
24 *Chile decide excluir a empresas israelíes de la FIDAE—DW—06/03/2024* (2024).
25 *InvestChile* (2024).

26 McClory and Harvey (2016), p. 2.
27 McClory and Harvey (2016), p. 5.
28 Jhee and Lee (2011), p. 52.
29 Weiss (2005).
30 Eriksson and Giacomello (2007), p. 5.
31 Eriksson and Newlove-Eriksson (2021).
32 Morgenthau (1948), p. 4.
33 Gellman (1988), p. 248.
34 Snyder (2002), pp. 151–152.
35 Waltz (1979), p. 73.
36 I use the word orthodox referring to heavily classical realists that dismiss the existence of other potential causes and factors for the phenomena they study outside of security and state self-preservation.
37 Keohane and Martin (1995), p. 42.
38 Eriksson and Giacomello (2007), p. 5.
39 Scheuerman (2009) refers to Global Warming or the effects of certain technologies that are seemly disregarded by main realist thinkers.
40 Direct quotation from Scheuerman (2009), online.
41 Scheuerman (2009).
42 Wendt (1995), pp. 71–72.
43 Wendt (1995), p. 73.
44 Haass (2020), pp. 160–161.
45 Wendt (1995), p. 73.
46 Wendt (1995), p. 74.
47 Copeland (2000), p. 188.
48 Moravcsik (1992), p. 1.
49 Steele (2007).
50 Moravcsik (1992), p. 5.
51 Moravcsik (1992), pp. 4–5.
52 Keohane and Martin (1995), pp. 39–41.
53 Keohane and Martin (1995), p. 42.
54 Moravcsik (1997), pp. 516–517.
55 Haass (2020), pp. 203–204; 297–299.
56 Higgott (2004).
57 Thomas (2017).
58 Thomas (2017).
59 Hurwitz (2014), p. 326.
60 Moravcsik (1997), p. 515.
61 Dunn Cavelty (2007), pp. 85–86.
62 Thomas (2017).
63 Yoshimatsu (2007).
64 Frenkel et al. (2020).
65 Rose (1998), p. 151; Gvalia et al. (2019), pp. 21–22.
66 Taliaferro et al. (2009), p. 3.
67 Taliaferro et al. (2009), p. 11.
68 Taliaferro et al. (2009), p. 22.
69 Ripsman (2009), pp. 170–173.
70 Coetzee and Hudson (2012), online.
71 Gourevitch (1978), p. 894.
72 Mueller et al. (2007), first page, electronic copy.
73 Gourevitch (1978), pp. 900–906.
74 Gourevitch (1978), p. 912.
75 Putnam (1988), p. 433.
76 Putnam (1988), pp. 435–437.
77 Foulon (2015), p. 640.

78 Vasileiadis (2023).
79 Edwards (2011), online.
80 Pearce (1997), pp. 64–66.
81 Pearce (1997), pp. 67–69.
82 Pearce (1997), p. 67.
83 Diamond (1994), pp. 5–6.
84 Rathbun (2008), p. 301.
85 Rathbun (2008), p. 316.
86 Putnam (1988), p. 434.
87 Lynn (2010).
88 Backes and Swab (2019).
89 Voon and Mitchell (2019).

Bibliography

Backes, O., & Swab, A. (2019). *Cognitive Warfare: The Russian Threat to Election Integrity in the Baltic States*. Belfer Center for Science and International Affairs. https://www.belfercenter.org/publication/cognitive-warfare-russian-threat-election-integrity-baltic-states

Berenskoetter, F., & Williams, M. J. (2007). *Power in World Politics*. Routledge.

Budapest Convention—Cybercrime. (n.d.). Retrieved February 27, 2024, from https://www.coe.int/en/web/cybercrime/the-budapest-convention

Chile decide excluir a empresas israelíes de la FIDAE – DW – 06/03/2024. (2024). dw.com. https://www.dw.com/es/chile-decide-excluir-a-empresas-israel%C3%ADes-de-la-fidae/a-68449856

Coetzee, E., & Hudson, H. (2012). Democratic Peace Theory and the Realist-Liberal Dichotomy: The Promise of Neoclassical Realism? *Politikon*, *39*(2), 257–277. https://doi.org/10.1080/02589346.2012.683942

Composite Index of National Capability by Country 2024. (2024). *World Population Review*. https://worldpopulationreview.com/country-rankings/composite-index-of-national-capability-by-country

Copeland, D. C. (2000). The Constructivist Challenge to Structural Realism: A Review Essay. *International Security*, *25*(2), 187–212. https://www.jstor.org/stable/2626757

De Soto, J. A. (2021). Evaluation of the Moderna, Pfizer/Biotech, Astrazeneca/Oxford and Sputnik V Vaccines for Covid-19. *Advance Research Journal of Medical Clinical and Science*, *7*(1). https://www.arjmcs.info/index.php/arjmcs/article/view/246

Diamond, L. (1994). Rethinking Civil Society: Toward Democratic Consolidation. *Journal of Democracy*, *5*(3), 4–17. https://doi.org/10.1353/jod.1994.0041

Dunn Cavelty, M. (2007). Securing the Digital Age: The Challenges of Complexity for Critical Infrastructure Protection and IR Theory. In Johan Eriksson & Giampiero Giacomello (Eds.), *International Relations and Security in the Digital Age* (pp. 85–105). Routledge. https://doi.org/10.4324/9780203964736

Edwards, M. (Ed.). (2011). Civil Society in Latin America. In *The Oxford Handbook of Civil Society* (1st ed.). Oxford University Press. https://doi.org/10.1093/oxfordhb/9780195398571.013.0010

Eriksson, J., & Giacomello, G. (Eds.). (2007). *International Relations and Security in the Digital Age*. Routledge. https://doi.org/10.4324/9780203964736

Eriksson, J., & Newlove-Eriksson, L. M. (2021). Chapter 1: Theorizing Technology and International Relations: Prevailing Perspectives and New Horizons. In Giampiero Giacomello, Francesco N. Moro, & Marco Valigi (Eds.), *Technology and International*

Relations. Edward Elgar. https://www.elgaronline.com/edcollchap/edcoll/9781788976060/9781788976060.00007.xml

Ferguson, N. (2012). *Civilization: The West and the Rest*. Penguin Books.

Foulon, M. (2015). Neoclassical Realism: Challengers and Bridging Identities. *International Studies Review*, *17*(4), 635–661. https://www.jstor.org/stable/24758570

Frenkel, A., Blinder, D., Frenkel, A., & Blinder, D. (2020). Geopolítica y cooperación espacial: China y América del Sur. *Desafíos*, *32*(1), 114–143. https://doi.org/10.12804/revistas.urosario.edu.co/desafios/a.7669

Gellman, P. (1988). Hans J. Morgenthau and the Legacy of Political Realism. *Review of International Studies*, *14*(4), 247–266. https://www.jstor.org/stable/20097151

Gerschenkron, A. (1962). *Economic Backwardness in Historical Perspective*. The Belknap Pr. of Harvard Univ. Pr; WorldCat.

Gilpin, R. (Ed.). (1981a). Growth and Expansion. In *War and Change in World Politics* (pp. 106–155). Cambridge University Press. https://doi.org/10.1017/CBO9780511664267.005

Gilpin, R. (Ed.). (1981b). The Nature of International Political Change. In *War and Change in World Politics* (pp. 9–49). Cambridge University Press. https://doi.org/10.1017/CBO9780511664267.003

Gourevitch, P. (1978). The Second Image Reversed: The International Sources of Domestic Politics. *International Organization*, *32*(4), 881–912. https://www.jstor.org/stable/2706180

Gunnarsson, C. (2016). Misinterpreting the East Asian Miracle—A Gerschenkronian Perspective on Substitution and Advantages of Backwardness in the Industrialization of Eastern Asia. In *Diverse Development Paths and Structural Transformation in the Escape from Poverty* (pp. 93–127). Oxford University Press. https://doi.org/10.1093/acprof:oso/9780198737407.003.0005

Gvalia, G., Lebanidze, B., & Siroky, D. S. (2019). Neoclassical Realism and Small States: Systemic Constraints and Domestic Filters in Georgia's Foreign Policy. *East European Politics*, *35*(1), 21–51. https://doi.org/10.1080/21599165.2019.1581066

Haass, R. (2020). *The World: A Brief Introduction*. Penguin Press. https://www.overdrive.com/search?q=CD753E1C-8E81-4B40-81A2-02883AE41A94

Heim, J., & Miller, B. (2020). *Measuring Power, Power Cycles, and the Risk of Great-Power War in the 21st Century*. RAND Corporation. https://doi.org/10.7249/RR2989

Higgott, R. (2004). US Foreign Policy and the 'Securitization' of Economic Globalization. *International Politics*, *41*(2), 147–175. https://doi.org/10.1057/palgrave.ip.8800073

Houweling, H. W., & Siccama, J. G. (1981). The Arms Race-war Relationship: Why Serious Disputes Matter*. *Contemporary Security Policy*, *2*(2), 157–197. https://doi.org/10.1080/01440388108403725

Hurwitz, R. (2014). The Play of States: Norms and Security in Cyberspace. *American Foreign Policy Interests*, *36*(5), 322–331. https://doi.org/10.1080/10803920.2014.969180

InvestChile. (n.d.). *FIDAE 2024*. Retrieved March 7, 2024, from https://www.investchile.gob.cl/fidae-2024/

Jhee, B., & Lee, N. (2011). Measuring Soft Power in East Asia: An Overview of Soft Power in East Asia on Affective and Normative Dimensions. In S. J. Lee, & J. Melissen (Eds.), *Public Diplomacy and Soft Power in East Asia* (pp. 51–64). Palgrave Macmillan US. https://doi.org/10.1057/9780230118447_4

Kania, E., & McCaslin, I. B. (2020). The PLA's Medical Contributions to COVID-19 Response. In *People's Warfare against Covid-19* (pp. 22–24). Institute for the Study of War. https://www.jstor.org/stable/resrep27524.8

Keohane, R. O., & Martin, L. L. (1995). The Promise of Institutionalist Theory. *International Security*, *20*(1), 39–51. https://doi.org/10.2307/2539214

Lynn, W. F. I. (2010). Defending a New Domain—The Pentagon's Cyberstrategy Essay. *Foreign Affairs*, *89*(5), 97–108. https://heinonline.org/HOL/P?h=hein.journals/fora89&i=803

Malacalza, B., & Fagaburu, D. (2022). ¿Empatía o cálculo? Un análisis crítico de la geopolítica de las vacunas en América Latina. *Foro Int. Ciudad de México*, *62*. https://www.scielo.org.mx/scielo.php?pid=S0185-013X2022000100005&script=sci_arttext

Malamud, C., & Núñez, R. (2021). *Vacunas sin integración y geopolítica en América Latina*. Real Instituto Elcano. https://www.realinstitutoelcano.org/analisis/vacunas-sin-integracion-y-geopolitica-en-america-latina/

McClory, J., & Harvey, O. (2016). The Soft Power 30: Getting to Grips with the Measurement Challenge. *Global Affairs*, *2*(3), 309–319. https://doi.org/10.1080/23340460.2016.1239379

Moravcsik, A. (1992). Liberalism and International Relations Theory. *Working Paper Series (Harvard University. Center for International Affairs)*. Paper No. 92-6, 1–38.

Moravcsik, A. (1997). Taking Preferences Seriously: A Liberal Theory of International Politics. *International Organization*, *51*(4), 513–553. https://www.jstor.org/stable/2703498

Morgenthau, H. (1948). *Politics among Nations* (1sr. (Digital)). A. A. Knoff.

Morgenthau, H. (1955). Politics among Nations: The Struggle for Power and Peace. By Hans J. Morgenthau. (New York: Alfred A. Knopf. 1954. 2nd ed. Pp. xxi, 600, xxv. $5.75.). *American Political Science Review*, *49*(2), 586–586. https://doi.org/10.1017/S0003055400275254

Mueller, M., Cogburn, D. L., Mathiason, J., & Hofmann, J. (2007). Net Neutrality as Global Principle for Internet Governance. In *GigaNet: Global Internet Governance Academic Network, Annual Symposium 2007*. https://doi.org/10.2139/ssrn.2798314

Nye, J. S. (2008). Public Diplomacy and Soft Power. *The Annals of the American Academy of Political and Social Science*, *616*(1), 94–109. https://doi.org/10.1177/0002716207311699

Pearce, J. (1997). Civil Society, the Market and Democracy in Latin America. *Democratization*, *4*(2), 57–83. https://doi.org/10.1080/13510349708403515

Powell, R. (2003). Nuclear Deterrence Theory, Nuclear Proliferation, and National Missile Defense. *International Security*, *27*(4), 86–118. https://www.jstor.org/stable/4137605

Putnam, R. D. (1988). Diplomacy and Domestic Politics: The Logic of Two-Level Games. *International Organization*, *42*(3), 427–460. https://www.jstor.org/stable/2706785

Rathbun, B. (2008). A Rose by Any Other Name: Neoclassical Realism as the Logical and Necessary Extension of Structural Realism. *Security Studies*, *17*(2), 294–321. https://doi.org/10.1080/09636410802098917

Ripsman, N. M. (2009). Neoclassical Realism and Domestic Interest Groups. In S. E. Lobell, N. M. Ripsman, & J. W. Taliaferro (Eds.), *Neoclassical Realism, the State, and Foreign Policy* (1st ed., pp. 170–193). Cambridge University Press. https://doi.org/10.1017/CBO9780511811869.006

Rose, G. (1998). Neoclassical Realism and Theories of Foreign Policy. *World Politics*, *51*(1), 144–172. https://www.jstor.org/stable/25054068

Scheuerman, W. E. (2009). Realism and the Critique of Technology. *Cambridge Review of International Affairs*, *22*(4), 563–584. https://doi.org/10.1080/09557570903325504

Skolnikoff, E. B. (1994). *The Elusive Transformation: Science, Technology, and the Evolution of International Politics*. Princeton University Press.

Snyder, G. H. (2002). Mearsheimer's World-Offensive Realism and the Struggle for Security: A Review Essay. *International Security*, *27*(1), 149–173. https://www.jstor.org/stable/3092155

Steele, B. J. (2007). Liberal-Idealism: A Constructivist Critique1. *International Studies Review*, *9*(1), 23–52. https://doi.org/10.1111/j.1468-2486.2007.00644.x

Taliaferro, J. W., Lobell, S. E., & Ripsman, N. M. (2009). Introduction: Neoclassical Realism, the State, and Foreign Policy. In S. E. Lobell, N. M. Ripsman, & J. W. Taliaferro (Eds.), *Neoclassical Realism, the State, and Foreign Policy* (1st ed., pp. 1–41). Cambridge University Press. https://doi.org/10.1017/CBO9780511811869.001

Thomas, E. (2017, November 13). *Taming the 'Wild West': The Role of International Norms in Cyberspace*. E-International Relations. https://www.e-ir.info/2017/11/13/taming-the-wild-west-the-role-of-international-norms-in-cyberspace/

Urdinez, F., & Winters, M. (2021). *COVID-19 Diplomacy and Soft Power: Did Vaccine and Equipment Distribution Improve Perceptions of China in Latin America?* https://www.furdinez.com/uploads/2/2/5/6/22565746/working_paper_-_urdinez___winters_2021_oct.pdf.

Vasileiadis, P. (2023). Reconstructing Neoclassical Realism: A Transitive Approach. *International Relations*, 00471178231185747. https://doi.org/10.1177/00471178231185747

Voon, T., & Mitchell, A. D. (2019). Australia's Huawei Ban Raises Difficult Questions for the WTO (SSRN Scholarly Paper 3390675). https://doi.org/10.2139/ssrn.3390675

Waltz, K. N. (1979). *Theory of International Politics*. Addison-Wesley Publishing Company.

Weiss, C. (2005). Science, Technology and International Relations. *Technology in Society*, *27*(3), 295–313. https://doi.org/10.1016/j.techsoc.2005.04.004

Wendt, A. (1995). Constructing International Politics. *International Security*, *20*(1), 71–81. https://doi.org/10.2307/2539217

Yoshimatsu, H. (2007). Global Competition and Technology Standards: Japan's Quest for Techno-Regionalism. *Journal of East Asian Studies*, *7*, 439–468. doi:10.1017/S1598240800002587

Zhao, J., Zhao, S., Ou, J., Zhang, J., Lan, W., Guan, W., Wu, X., Yan, Y., Zhao, W., Wu, J., Chodosh, J., & Zhang, Q. (2020). COVID-19: Coronavirus Vaccine Development Updates. *Frontiers in Immunology*, *11*. https://www.frontiersin.org/journals/immunology/articles/10.3389/fimmu.2020.602256

5 Reasons Behind Decision-Making in International Technology Transfers

So far, this book has introduced the issue of technology transfers amidst the increasing rivalry between the US and the PRC. It discussed philosophical and sociological views on the importance of innovation, technology's relevance to international relations, how theory approaches technology creation, acquisition, adoption, and transfers—and how it narrows down to the garnering of power through the capacity to coerce and persuade. In this chapter, I explore the incentives and motivations of countries to receive and transfer technology: economic incentives, values and culture, and security and geopolitical considerations. When referring to Latin American cases, more than one of these variables may come into play, and I consider this and reflect it in the case study's chapters. The aim is, however, to frame and simplify these concepts enough so that a clear link can be established between the results of these processes and the variables mentioned above.

Theorizing on Technology Adoption

Who adopts a new technology? Is it individuals, organizations, or states? I ask this question from a practical and philosophical perspective since characteristics determine how fast a unit can identify and value innovation and where concerns and lack of skills and tools limit the adoption. In that regard, extra-disciplinary research can help frame both aspects. One of the groundbreaking authors in this area, Everett Rogers, identified "types" of adopters and "stages." The "types" range from innovators, early adopters, early majority, late majority, and laggards,[1] whereas the stages are five clear stages: knowledge, persuasion, decision, implementation, and confirmation.[2] For him, this decision process aims at advancing the level of information an actor has motivated to reduce uncertainty and measure innovation's pros and cons.[3] Innovation requires a system that supports the production of software and hardware—directly related to what has been discussed in previous chapters—and has an inherent level of uncertainty connected to the adoption by other actors.

While many companies might successfully innovate and create solutions, not all will successfully disseminate those technologies. Standardization serves economic, security, and social objectives. It reduces costs, increases transparency, and boosts efficiency across markets.[4] In other words, the massification of technology depends not only on the product itself but also on the political and practical

DOI: 10.4324/9781003489450-6

qualities associated with managing and marketing it. Likewise, the acquisitions of innovation might not have access to all available options, or these cannot reach them for strategic reasons. This was evident during the Cold War in cases where Western and Soviet companies could not readily cross the so-called "Iron Curtain," limiting options and access to certain types of innovation.

To correctly identify reasons behind the adoption, transfer, and diffusion of technology, it is necessary to assess elements that include practical outcomes analysis, relative advantages, complexity, compatibility, and so on, as well as communication processes that generate reactions in the technology diffusion process.[5] Even more; opinion leadership is a critical element in the diffusion of technologies.[6] For states, geopolitical, geographical, institutional, social, or even historical circumstances explain technological transfer, especially between major powers such as the US.[7] Therefore, the standardization of innovation results from a multitude of factors, including communication among stakeholders, the specific strategies of the private sector's actors in a competitive scenario, and the regulation from governmental institutions. The nationwide implementation of technology, even more so one created in a foreign state, relies on the private sector but is heavily overseen by the state.[8]

This creates a picture of what the process looks like. Decision-makers assess a technology, which—clouded by uncertainty—will have both positive and negative consequences that are generally relevant to their environment. To do this, they consider possible advantages and outcomes and how compatible said technology is with what they already have. How quickly they implement existing technology compared to other actors will determine where they are located in the overall characterization of adaptors. An example of this is digital payments. For cash-based societies, facilitating card and mobile payments represents state and individual decisions. Governments must consider security and systemic considerations when putting forward legislation facilitating the entrance of hardware and software providers. The evolution of banking systems and the traceability of money are both immense debate points for policymakers. On the other hand, even though the legal and service frameworks allow for cashless transactions, individuals might take a while to adapt and change to a new system. As a result, some countries will take longer to implement cashless alternatives widely than others.

Latin American governments have long been active in spreading the adoption of innovation in their jurisdictions, spending part of their budget to modernize and propose highly complex projects in terms of infrastructure and IT.[9] The justification for their active role in promoting technological adaptation is the capability for economic and social development, which is also typically endorsed by regional entities such as ECLAC or the OAS, and as a result of this regional philosophy, no sooner had the internet found a foot in globalization processes—in the year 2000—than several American states agreed to this developmental role in communication and information technologies.[10] The objectives were relatively homogeneous across borders: ensure equality in connectivity, preserve a neutral and open market for competition, and digitalization of government procedures and societies, with digital literacy at the forefront of IT dissemination.

This means that these countries are actively spreading these foreign technologies, and it is expected, for most relevant cases such as 5G, that the administrations will have an active role in their relationship with the decision-making process. Of course, this also includes the financing of these processes. While European and American companies were predominant in the 1990s and early 2000s, China has recently become a dominant actor, taking advantage of this active seeking for progress and infrastructure gaps in their strategy toward the region, and this explains their rapid growth in the IT sector.[11] Even though other chapters in this book will elaborate on the stated tensions between the US as a force that has led the region for more than a century, and China as a rising power that defies both the status quo of the region and the values of Western democracies, it is essential to understand that Latin American states take action in a complex environment that balances the urge for neutrality and the inherit deontological nature of democratization, which transpires in regional agreements and also international standards.

I propose that all these elements are coherent with a neoclassical realist approach, as it can be argued that an individual's approach to innovation, especially decision-makers, will permeate the overall output of the technology being implemented into a national system. In this interaction, international and narrative factors might interfere with the decisions. For this book, decision-makers who push to adopt these technologies are mainly policymakers. Still, the role of the academia, the media, and the private sector's perspectives are also incorporated into the analysis.

In the context of case studies relevant to Latin America, countries consider three lines of justification for choosing a specific provider or side with a particular standard. The first is, of course, economic. Countries will select a technology or a standard according to their financial capacity and an economically savvy strategy in the long term. The second one is geopolitical. In an increasingly complex international scenario, decisions that involve the influence spheres of powers have repercussions that impact the region in waves. The third one is shared values and norms. Regulations reflect specific value-related frameworks that might not be universally shared, and soft power is utilized to gain attraction and shape decisions.

The Economic Reasons behind Technology Adoption

Latin American countries often experience economic crises and financial vulnerability and are in the spotlight of repeated corruption and embezzlement connected to infrastructure projects.[12] As a result, there is a clear correlation between what these governments are willing or can afford regarding the type of offerors and the results of these processes. A cheaper price and higher returns are always attractive to decision-makers. Nevertheless, price is not the only guiding factor when a state makes nationwide decisions in innovation.

For example, financial incentives, such as loans or grants or direct negotiation between governments in which competitors are isolated from participating, are also possible elements that will affect a country's behavior, especially considering countries that do not have access to international financial markets.[13]

Suppose economic incentives are the main driving force behind technological transfer. In that case, decision-makers will likely base their choice on the economic situation and financial proposal of the available options. Meanwhile, technology providers are incentivized to spread their standards globally to enhance their economic competitiveness,[14] in a process called "harmonization" of technology. This has two characteristics. On the one hand, it increases a company's or actor's participation in a market when it limits the user's options to adopt different standards.

On the other hand, harmonization of standards creates economic opportunities, helps reduce costs, and increases efficiency.[15] This process is also contagious.[16] Others tend to follow when a regional agency, leader, or relevant actor is faster in implementing a novel technology. States have prioritized this option since the multilateral trade institutions pushed countries away from closed systems.[17] Harmonization occurs for several reasons. When technology is a novelty and few actors are involved, states might choose one provider for reasons ranging from random assessment to previous history. That provider is then recommended in bilateral and multilateral instances, which creates a regional standard. In addition, there might also be regulations that enforce or limit the number and characteristics of participants in a system. Portugal-Perez et al. make an apparent distinction between these two forms of harmonization:

> Product standards and technical regulations set out product characteristics, related processes, and production methods. The World Trade Organization (WTO) differentiates them by their compliance degree: a technical regulation is mandatory, while a product standard is voluntary. In practice, this distinction is blurred, as public agencies often use standards to achieve regulatory goals. Standards affecting electronic products cover a wide range of product specifications.[18]

As I mentioned before, harmonization is also attractive for providers. When a region is accustomed to a certain offeror, companies have an advantage even if they don't offer the most economically savvy option. Yoshimatsu also states that when a state or a region creates and follows standards, the whole system gains an advantage in innovation and diffusion.

> [Standards] are valuable in industrial sectors where network externalities matter. (…) These network externalities, however, require compatibility and interoperability in the exchange of data and products and raise the importance of the broader diffusion of standards that cover such interoperability and compatibility. Technology standards also provide substantial benefits for consumers by bolstering the credibility of technologies (…). Even if a firm invents an innovative technology (…) that firm or industry cannot become a winner if it fails to make the technologies accepted in the regional or global market. Firms have to transition from their own to international standards. The government should have strong incentives

> to encourage firms to make efforts to diffuse technology standards in the marketplace.[19]

In that sense, it is sensible to say that companies—and states that back companies within their public diplomacy—look forward to providing innovation that can become standardized abroad or that is preferred as the regional standard. This is not only an economic advantage but also a specific differentiation in each country's offer. Moreover, countries can be expected to apply "economic" and "commercial" diplomacy to spread these standards further. The reason countries use their institutions to support national companies, even if they are private entities.[20]

This is also enhanced by the interaction between government ministries (such as MOFAs) and international institutions directly related to foreign economic affairs and globalized institutions such as the International Monetary Fund (IMF), the OECD, and the WTO.[21]

Of course, private companies in democratic countries enjoy relative independence and the capacity to act following their expected benefits. Although not a definitive rule, on many occasions, their interaction with their base country's institutions is entangled with the specific interests shared between actors. Therefore, this is an indirect relationship in which companies seek the help of their governments within international institutions.[22]

Latin America has had mixed behaviors in harmonization related to economic incentives. For some countries, technical considerations stand before regional preferences or price, whereas others are influenced by externalities, collaboration opportunities, and political relationships with their neighbors.[23] An important example of this, which I intend to deepen in Chapter 6, is the arrival of satellite television in Latin America. The Japanese alternative ISDB-IT was shown to be preferred in the South American Region over the American (ATSC), Chinese (DMB-T), and European (DVB-T), and the reason for this is mainly related to Brazil's early decision-making process.[24]

Limiting the Impact of Low Offers in Competition

Individual price is not the only factor to consider when assessing the economic element of a country's decision-making. Moreover, some countries and organizations have implemented rules to avoid unlawful or unrealistically low offerings, especially in public tenders or auctions. For example, the Agreement on Government Procurement (GPA) within the WTO established that:

> Where a procuring entity receives a tender with a price that is abnormally lower than the prices in other tenders submitted, it may verify with the supplier that it satisfies the conditions for participation and is capable of fulfilling the terms of the contract.[25]

Nevertheless, no Latin American country is currently a party to the WTO agreement. Brazil applied for accession in May 2020, while countries like Argentina, Chile,

Colombia, and Ecuador act as observers but are not negotiating accession or a GPA commitment before application.[26]

In any case, the provision expressed in the system of countries that abide by this treaty restricts aggressive economic competition, which might play havoc with quality and safety. In addition, some countries will accept "reckless" offers even if they are way under the price of the other competitors when the justification for those processes is properly assessed.[27] In Europe—and specifically Spain—an offer will not be considered too low until the process to determine the causes of said price is presented before the institution and said procurement institution decides whether to incorporate the offer or not, based on the capacity of the company or external factors.[28]

Economic Standards and Value-Sharing Standards

International norms are followed by countries not because they can be enforced in a traditional sense—since there isn't a global monopoly of violence—but because they are considered valid and hold legitimacy. These norms, which can take a plethora of shapes, not only offer guidelines for behavior but also carry a moral judgment and can be used to criticize or approve specific actions.[29] Technological standards and related norms can also be guided and interpreted from an international scope.[30]

When a country adopts an innovation that requires infrastructure or specific hardware, decision-makers in the private and public sectors are driven to set rules and standards to enable its use for the target population. Even though the use of technology is private or personal in citizens' lives, there is a need for efficiency, interoperability, and compliance with international norms.[31]

Since 1990, standards have been implemented due to (a) racing toward dissemination, (b) economic benefits and harmonization, and (c) security concerns.[32] In that context, I divide standards into economy and value-oriented categories. The economy-oriented ones have to do with the harmonization and procedural considerations previously discussed. In contrast, the value-oriented ones are related to legal and behavioral frameworks that are not directly focused on economic sustainability. In any case, both sometimes interact, for example, in screening policies.[33] Both categories are constantly being challenged by aspiring rule-setters who wish to get more relative power in terms of political power dynamics or market share/presence.[34]

Reasons Based on Values and Norms in Decision-Making

Values are inherently abstract and can be challenging to define precisely in terminological contexts. Consequently, shared values present a notable challenge in international assessments. Values are closely intertwined with ideologies, encapsulating ideas and worldviews that inform behavior and culture.

Value affinity toward science and technology has been addressed by researchers, especially on a personal and psychological level. Citizens have been shown to exhibit a wide range of behaviors, from technophobes (fear of technology) to technophiles (love for technology), with these behaviors influencing innovation, development, and government outcomes at a societal level.[35]

Certain delimitations are helpful in comparing states' positions on value-related decision-making. A country's level of democracy, measured by several indexes, gives an idea of the state's priorities and values. An authoritarian regime may prioritize government survival over civil rights, while full democracies operate under legal and control-based frameworks. This is translated into the expectations leaders hold for other rulers.

In technology-related policy, some common discussions are closely related to values that "may include liberty, justice, enlightenment, privacy, security, friendship, comfort, trust, autonomy, and sustenance."[36] For some actors and researchers, technological standards are—and should—always be linked to protecting and respecting the value systems of democratic societies. Hurwitz, on the other hand, states that cyber powers had been agreeing on cyber-related issues with minimum conventions, and the inclusion of liberal democracy's elements that were meant to be pushed internationally has somewhat fragmented and accelerated the process of norm-challenging.[37] The waves of the fragmentation process reach the Latin American regions. In the face of offers that present themselves as the reflection of a set of values, it is not enough to comply with the international norms, and states take a decision that implies they will reflect one ideological framework—liberal—or the alternatives.[38] As Vázquez Callo-Müller describes:

> Current privacy laws that regulate data collection, storage, and processing are being adopted on a national or regional level (this is the case of the European Union), leading to the regulatory fragmentation of a very global phenomenon. This raises questions of interoperability across national data protection regulations. (…) it has been argued that in some cases, differences in data protection laws can shift comparative advantage to certain locations.[39]

Within that fragmented context, some authoritarian actors have pushed for new concepts to challenge the general conventions, such as sovereignty and security. For example, internet standards or parallel standards, as is the case of China and Russia within the Budapest Convention on Cybercrime, to achieve a more "state-controlled governance model." Russia did this more directly by campaigning for the secretary general position in the International Telecommunications Union (ITU) against the US candidate.[40] The elected American candidate, Doreen Bogdan-Martin, took office in early 2023. Nevertheless, it is imperative to note that this contested scenario is anticipated to endure in the foreseeable future. The Chinese government is also in a race to establish internet sovereignty as a legal norm, which could undermine transparency, accountability, and human rights commitments,[41] and attempts to do this by extending its cyber borders, which translates to the transfer of internet, clouding, and wireless technology infrastructure; public services as well as financial, telecommunications, transportation, and other networked systems.[42]

Because of the same fragmentation process, the issue of like-mindedness has become a topic of discussion, not only in international agoras but also domestically. The World Values Survey (WVS), a renowned source for measuring attitudes across countries, provides a solid glimpse into the thoughts of individuals on specific aspects of technology and the accepted limits of government interference in citizens' privacy. When asked if the government has the right to collect information about anyone without their knowledge, Latin American countries are not homogenous; some call for privacy against surveillance, and some show more outstanding shares of the population who prefer security over freedom.[43] However, it is common practice in the region to replicate or adopt legislation that prevents discrimination and violations of human rights when implementing technologies that need to collect personal information.[44] Privacy laws in Latin America, for example, are closely based on the concept of *habeas data*[45]—but this is not the case for those that have experienced a decrease in democratic standards in the last years, such as Venezuela or Nicaragua.

Moreover, the Coronavirus pandemic increased the tense relations between privacy and security.[46] There is also little homogeneity regarding the region's relationship and trust in science. According to the last WVS, Ecuador and Colombia remain skeptical over the role of science and technology in their life. One could say that citizens value technology in a very different way across the region and compared to countries such as the US, China, or Japan.

The Question of Value and Valueless Technology

Whether it is possible to judge technology based on values that could be embedded into artifacts or systems is an ongoing discussion.[47] Currents of thought assume that technology should not have a value-based analysis, as its ethics derive from usage rather than capability. A rocket can propel a missile or spacecraft, the internet began as a military function before becoming a civilian tool, and social media can trigger social control and political change. In other words, there is agreement, to a sensible extent, that usage is what provides a technology with an ethical framework.

Nevertheless, there are instances when the dual or neutrality of innovation could be more apparent, such as with weapons that are created for mass destruction.[48] Technologies can be technically valuable in terms of efficiency and functionality and also have a social, moral, and political added value—or hindrance—to society that makes those who create said technology responsible for their outcomes.[49] Cultural and political contexts can shape these values differently, and therefore, influence the ethical implications of these technologies for the end users.[50]

This is why technical law exists: to enable technology and to limit its possible adverse outcomes.[51] In this context, the law should be a medium to imprint the values of constitutional states and steer the use of technology and, hence, its behavior and developers, understanding the complexity and incapacity to control all aspects of the creation of inventions.[52]

Values and the Measurement of Soft Power

Never mind the inherent objectives of a provided technology, I argue that perception is—as in many cases in international relations—more critical than actual capacity. The quality and background of any provider are often related to governments' perception of a foreign state. In other words, there is bidirectional feedback between companies or producers and their country of origin regarding technology quality, prestige, or thoroughness.[53] Personal decision-making provides an excellent example of this. Imagine we are deciding to buy a car and choose a German one over a hypothetical Chilean vehicle. In that case, do I consider German car brands reliable because they are German technology built under a determined culture and system? Or do we believe that German technology is trustworthy because we have gotten used to German cars being good due to their private practices over time? The answer should be both. A Chilean car might be as good as a German one in the long term if the right business culture and know-how are correctly implemented. Still, the public trust will be built over a considerably extended period. The assumption is that a prestigious technology will be associated with characteristics associated with the country, including abstract values such as freedom, trust, and commitment.

Because soft power aims at influencing behavior from other actors not by force or economic cohesion but using attraction,[54] the perception that decision-makers and the general population have of the technology provider will be a crucial element when leaning toward an alternative.[55]

Taking Security into Consideration When Making Decisions

Defining "security" can be as ambiguous as values in this context because it is a broad concept constantly revisited and reaches the military and the economic, humanitarian,[56] cyber, and other realms. It is mainly connected with preserving and protecting something from an external threat that can take multiple forms. Therefore, international alliances, states, organizations, and individuals assess and can feel threatened in the military sense and in any way that will affect their integrity. In human security, for example, this could include an event that disrupts food supplies or a disease that affects the population, even if there isn't an identifiable and accountable actor causing said disruption.

[Cyber] security concerns have been addressed as a primary driver for the technological adaptation landscape in Latin America, particularly within the context of a rivalry between the US and the PRC.[57] Directly related is the resurgence of geopolitics and the region's role in navigating the uncertain and complex scenario in which both forces represent critical economic partners and investors—a key motivator for state actions.[58] Due to this geopolitical landscape, some states might make decisions not for their financial soundness or value-related issues but for more politically driven factors such as power—the aforementioned hard power and smart power—alliances and the direct actions of the incumbent actors.[59]

Decision-makers can assess geopolitical strategic actions and security concerns when implementing a standard or technology. In fact, technology standards—understood from a modern take on the international order—have generally been addressed from a security point of view. In warfare, this has been critical following international conflicts. Poisonous gases, white phosphorus, and nuclear weapons serve as illustrative examples of innovations subject to stringent limitations or outright prohibition. More recently, integrating automated weapons, drones, and artificial intelligence into warfare has presented a complex ethical and regulatory dilemma.[60]

However, security in terms of more commonly used technologies, like those involved in telecommunications and infrastructure, is increasingly essential.[61] This stems—as in all other forms of innovation—from the potential geopolitical ramifications and the clash of values associated with both the application of these technologies and their societal repercussions.[62] States are incentivized to shape and restrict technologies and to expand or constrain their influence. Furthermore, when multiple entities vie for contracts through public tenders, specific attributes are often preferred in the companies and entities involved in these processes.

This is also true for Latin America. First, data privacy and cybersecurity breaches have increased in the past years across countries in the region.[63] Second, since governments are looking at international trends, they see other states' behavior as a probable guide for their course of action. A clear example is the European GDPR, often referred to when discussing amendments to national data privacy regulations. Third, the political, academic, and legal analysis of the issue transpires to states' reactions in the face of innovations with significant social impact.

The Ambiguity of Private and Public Actors

In this book, I analyze the discourse surrounding space-related initiatives, identity recognition systems, the deployment of 5G networks, and other telecommunication technologies necessitating substantial investments. The objective is to ascertain whether states are concerned about security implications throughout these endeavors. Additionally, the examination aims to discern whether such considerations primarily prioritize safeguarding the general population or if discourse surrounding sovereignty has predominantly overshadowed these deliberations. The main problem when comparing potential security threats in technology investments is the dominant role of private actors, and the arguable control states have of companies abroad. In other words, the states in question also need to assess an ambiguous relationship with domestic and foreign companies. According to an OECD report on the matter:

> In the coming years, the sector may therefore see an increase in collaboration between networks and vertical industries. New partnerships are arising, not only among industry verticals and horizontal players but also among countries. In the European Union, a clear example is the 5G corridors (i.e. highways)

that involve the collaboration of many European countries to prepare for a future with fully automated vehicles that may potentially use 5G.[64]

Throughout 2017 and 2018, numerous European governments—including the "Big Three" recipients of Chinese capital, Germany, France, and the United Kingdom (UK)—have proposed or passed new legislation that increases the scrutiny of foreign mergers and acquisitions for potential national security risks.[65] Regarding the Australian ban on Huawei, a security exception argument has been raised to shield the decision from international law, mainly related to the WTO jurisdiction.[66] This has been criticized, but it is a clear and straightforward example of how domestic and global notions of security interact. It also addresses perceptions of threats from the ambiguous relationship between China and its companies. In the Australian case, "the evidence (…) suggests that political rhetoric encourages domestic audiences to engage in debates that have benign outputs for security."[67] In that sense, the discussion over securitization among the public is related to the country's policy. Convincing and engaging the audience with the country's security provisions would be critical when explaining how the policy has been shaped.[68]

Identifying Rationale and Actors

The interaction between private and public actors and the complex interconnectivity and presence of different layers in which negotiations are made make it sometimes difficult to identify one sole line of causation between the rationale a state follows to choose a provider and the preferred provider. In some cases, decision-makers might consider and weigh all potential factors before deciding. In others—particularly in Latin America's infrastructure arena—extra-official incentives might play havoc with these processes' results. Cases like Odebrecht and Lavagiato are examples of how corruption is embedded in Latin American politics and crosses frontiers to influence outcomes regionally. In other cases, the lack of funds and attractive markets jeopardizes a state's capacity to attract a wide range of competitors, which is utilized as an advantage by actors with power and political interests in the region. After all, this type of process only works if companies interact and compete to implement a technological system or service. I will discuss this later when reviewing facial recognition technology and cloud services. Finally, besides traditional corruption, states sometimes engage in secret contracts with governments without open competition and do not disclose the reasons and conditions of their interests. In a later chapter, I describe the Neuquén (Argentina) deep space observatory, which has been discovered to be directly operated by the People's Liberation Army and has alternate operations apart from those it is said to conduct.

Needless to say, in Latin America, the US holds a permanent political presence, but China is seen as a rising competitor. Some authors argue that China has not altered the Latin American institutional dynamic, but this is contested by those who express that both the US and the PRC pressure the region regarding technological implementation.[69] In the following part of this book, I discuss this dynamic and the actual influence both states have in terms of economic, political, and technological presence. I do this by presenting several cases of technology transfer and competition across Latin America in which the US, China, and other actors are competing for dominance. I then assess the reactions and decisions of the technology-receiving states to understand the rationale behind these transfers' outcomes.

Notes

1 Rogers (1983), pp. 247–250.
2 Rogers (1983), p. 165.
3 Rogers (1983), p. 163.
4 (Ergas, 1987), p. 192.
5 Rogers (1983), pp. 13–17.
6 Rogers (1983), p. 27.
7 Seely (2003), pp. 7–9.
8 Lopes de Souza (2009).
9 Ponce Regalado and Rojas Sifuentes (2010), p. 2.
10 Ponce Regalado and Rojas Sifuentes (2010), pp. 4–7.
11 Dongzhen et al. (2018), p. 23.
12 Latin America has been sadly known internationally for ist management of infrastructure-related corruption scandals such as the Odebrecht case (Campos et al., 2021).
13 Balbo and Cesarín (2020), p. 48.
14 Yoshimatsu (2007), p. 442.
15 Portugal-Perez et al. (2009), p. 2.
16 Angulo et al. (2011), p. 778.
17 Yoshimatsu (2007); Portugal-Perez et al. (2009).
18 Portugal-Perez et al. (2009), p. 18.
19 Yoshimatsu (2007), p. 442.
20 Saner and Yiu (2003), p. 3.
21 Saner and Yiu (2003), p. 5.
22 White (2015).
23 Angulo et al. (2011), pp. 773–775.
24 Angulo et al. (2011), p. 779.
25 WTO (1994), *Art. XV. 6.*
26 *WTO| Government Procurement—The Plurilateral Agreement on Government Procurement (GPA).*
27 Castel Aznar et al. (2021), p. 541.
28 Castel Aznar et al. (2021), pp. 542–544.
29 Thomas (2017).
30 Dunn Cavelty (2007), p. 107; Angulo et al. (2011); Yoshimatsu (2007); Hurwitz (2014); McKune and Ahmed (2018), p. 3850; Thomas (2017).
31 Hurwitz (2014), p. 332.
32 *Cybersecurity Developments in Latin America* (2018).
33 Hanemann et al. (2018), p. 9.
34 Yoshimatsu (2007).
35 Boehme-Neßler (2011: 1), p. 1.
36 Flanagan et al. (2008), p. 322.
37 Hurwitz (2014), pp. 323–326.
38 Vázquez Callo-Müller (2020), p. 9.
39 Vázquez Callo-Müller (2020), p. 9.
40 Sherman (2022), online.
41 McKune and Ahmed (2018), p. 3836.
42 McKune and Ahmed (2018), p. 3837.
43 Haerpfer et al. (2022), question 198.
44 Madsen (1998), p. 10; Brian Nougrères (2016), p. 147.
45 Kobek and Caldera (2016).

46 McNicholas et al. (2020), online.
47 Flanagan et al. (2008), p. 322.
48 Boehme-Neßler (2011: 1), p. 5.
49 Flanagan et al. (2008), p. 322.
50 Flanagan et al. (2008), p. 328.
51 Boehme-Neßler (2011: 1), p. 10.
52 Boehme-Neßler (2020).
53 McClory and Harvey (2016), p. 2.
54 Nye (2008), p. 94.
55 Jhee and Lee (2011), p. 52.
56 Rothschild (1995), p. 56.
57 Aguilar Antonio (2021); Colombo et al. (2021); Lee (2020).
58 Hanemann et al. (2019); Tekir (2019).
59 Balbo and Cesarín (2020).
60 Anderson and Waxman (2017).
61 Eriksson and Giacomello (2007), p. 5.
62 McKune and Ahmed (2018), p. 3850.
63 Aguilar Antonio (2021).
64 OECD (2019), p. 64.
65 Hanemann et al. (2018), p. 9.
66 Voon and Mitchell (2019).
67 McLean (2016).
68 McLean (2016).
69 Balbo and Cesarín (2020), p. 25.

Bibliography

Aguilar Antonio, J. M. (2021). Retos y oportunidades en materia de ciberseguridad de América Latina frente al contexto global de ciberamenazas a la seguridad nacional y política exterior. *Estudios Internacionales*, *53*(198), Article 198. https://doi.org/10.5354/0719-3769.2021.57067

Anderson, K., & Waxman, M. C. (2017). Debating Autonomous Weapon Systems, Their Ethics, and Their Regulation under International Law. In R. Brownsword, E. Scotford, & K. Yeung (Eds.), *The Oxford Handbook of Law, Regulation and Technology* (Vol. 1, pp. 1097–1117). Oxford University Press. https://doi.org/10.1093/oxfordhb/9780199680832.013.33

Angulo, J., Calzada, J., & Estruch, A. (2011). Selection of Standards for Digital Television: The Battle for Latin America. *Telecommunications Policy*, *35*(8), 773–787. https://doi.org/10.1016/j.telpol.2011.07.007

Balbo, G., & Cesarín, S. M. (2020). *¿Guerra comercial, periferia tecnológica o tecnoimperialismo?:* América Latina ante la competencia global en el sector de las telecomunicaciones. ALADI. https://repositorio.aladi.org/handle/20.500.12909/30827

Boehme-Neßler, V. (2011). Caught between Technophilia and Technophobia: Culture, Technology and the Law. In V. Boehme-Neßler (Ed.), *Pictorial Law: Modern Law and the Power of Pictures* (pp. 1–18). Springer. https://doi.org/10.1007/978-3-642-11889-0_1

Boehme-Neßler, V. (2020). Algo-democracy: Power of Technology, Powerlessness of Democracy? In V. Boehme-Neßler (Ed.), *Digitising Democracy* (pp. 61–95). Springer International Publishing. https://doi.org/10.1007/978-3-030-34556-3_4

Brian Nougrères, A. (2016). Data Protection and Enforcement in Latin America and in Uruguay. In D. Wright, & P. De Hert (Eds.), *Enforcing Privacy* (Vol. 25, pp. 145–180). Springer International Publishing. https://doi.org/10.1007/978-3-319-25047-2_7

Campos, N., Engel, E., Fischer, R. D., & Galetovic, A. (2021). The Ways of Corruption in Infrastructure: Lessons from the Odebrecht Case. *Journal of Economic Perspectives*, *35*(2), 171–190. https://doi.org/10.1257/jep.35.2.171

Castel Aznar, L., Montalbán Domingo, L., Pellicer Armiñana, E., & Catalá Alís, J. (2021). Abnormally Low Tenders in Public-Works Procurement: Determining Factors and Containment Measures. *25th International Congress on Project Management and Engineering*. http://dspace.aeipro.com/xmlui/bitstream/handle/123456789/2919/AT02-021_21.pdf?sequence=1&isAllowed=y

Colombo, S., López, M. P., & Vera, N. (2021). Tecnologías emergentes, poderes en competencia y regiones en disputa: América latina y el 5G en la contienda tecnológica entre China y Estados Unidos. *Estudos Internacionais: revista de relações internacionais da PUC Minas*, *9*(1), Article 1. http://hdl.handle.net/11449/206412

Dongzhen, Y., Jingsheng, D., Jianmin, Y., Zhenxing, S., Shixue, J., Yinghua, Z., Fan, Z., Baiyi, W., Haibin, N., Xiaodai, X., Shoujun, C., Zheng, Z., Cunhai, G., & Shuangrong, H. (2018). *Pensamiento social chino sobre América Latina* (W. Baiyi, Ed.). CLACSO. https://doi.org/10.2307/j.ctvnp0jw3

Dunn Cavelty, M. (2007). Securing the Digital Age: The Challenges of Complexity for Critical Infrastructure Protection and IR Theory. In *International Relations and Security in the Digital Age* (pp. 85–105). Routledge. https://doi.org/10.4324/9780203964736

Ergas, H. (1987). Does Technology Policy Matter. In *Technology and Global Industry: Companies and Nations in the World Economy. Center for European Policy Studies*, (pp. 4–69). National Academies Press.

Eriksson, J., & Giacomello, G. (Eds.). (2007). *International Relations and Security in the Digital Age*. Routledge. https://doi.org/10.4324/9780203964736

Flanagan, M., Howe, D. C., & Nissenbaum, H. (2008). Embodying Values in Technology: Theory and Practice. In J. Van Den Hoven, & J. Weckert (Eds.), *Information Technology and Moral Philosophy* (1st ed., pp. 322–353). Cambridge University Press. https://doi.org/10.1017/CBO9780511498725.017

Haerpfer, C., Inglehart, R., Moreno, A., Welzel, C., Kizilova, K., Diez-Medrano, J., Lagos, M., Norris, P., Ponarin, E., & Puranen, B. (2022). *World Values Survey Wave 7 (2017–2022) Cross-National Data-Set* (Version 4.0.0) [Dataset]. [object Object]. https://doi.org/10.14281/18241.18

Hanemann, T., Huotari, M., & Kratz, A. (2018). *Chinese FDI in Europe: 2018 Trends and Impact of New Screening Policies*. Merics & Rhodium Group (Updated in 2019).

Hurwitz, R. (2014). The Play of States: Norms and Security in Cyberspace. *American Foreign Policy Interests*, *36*(5), 322–331. https://doi.org/10.1080/10803920.2014.969180

Jhee, B., & Lee, N. (2011). Measuring Soft Power in East Asia: An Overview of Soft Power in East Asia on Affective and Normative Dimensions. In S. J. Lee, & J. Melissen (Eds.), *Public Diplomacy and Soft Power in East Asia* (pp. 51–64). Palgrave Macmillan US. https://doi.org/10.1057/9780230118447_4

Kobek, L. P., & Caldera, E. (2016). Cyber Security and Habeas Data: The Latin American Response to Information Security and Data Protection. *Oasis*, *24*, 109–128. https://www.redalyc.org/journal/531/53163716007/html/

Lee, N. T. (2020). *Navigating the U.S.-China 5G Competition*. Global China, Brookings. https://www.brookings.edu/articles/navigating-the-us-china-5g-competition/

Lopes de Souza, T., & Saboia Lima de Souza, R. (2009). Building the Digital TV Standard: The Brazilian Experience. *2009 GLOBELICS Conference*. https://citeseerx.ist.psu.edu/document?repid=rep1&type=pdf&doi=f824f80e0862fb26f1372089bf8ec58da71dde32

Madsen, W. (1998). Latin America May Follow European Privacy Lead. *Computer Fraud & Security*, *1998*(12), 10. https://doi.org/10.1016/S1361-3723(98)80037-7

McClory, J., & Harvey, O. (2016). The Soft Power 30: Getting to Grips with the Measurement Challenge. *Global Affairs*, *2*(3), 309–319. https://doi.org/10.1080/23340460.2016.1239379

McKune, S., & Ahmed, S. (2018). Authoritarian Practices in the Digital Age| The Contestation and Shaping of Cyber Norms through China's Internet Sovereignty Agenda. *International Journal of Communication, 12*. https://ijoc.org/index.php/ijoc/article/view/8540

McLean, W. (2016). Neoclassical Realism and Australian Foreign Policy: Understanding How Security Elites Frame Domestic Discourses. *Global Change, Peace & Security, 28*(1), 1–15. https://doi.org/10.1080/14781158.2015.1112774

McNicholas, E. R., Berg, N. M., Calvet, M. G., & Siveski, Nataša. (2020). *Data Privacy Concerns for Latin American Businesses during COVID-19 | Insights*. Ropes & Gray LLP. https://www.ropesgray.com/en/insights/alerts/2020/06/data-privacy-concerns-for-latin-american-businesses-during-covid-19

Nye, J. S. (2008). Public Diplomacy and Soft Power. *The Annals of the American Academy of Political and Social Science, 616*(1), 94–109. https://doi.org/10.1177/0002716207311699

OECD. (2019). *The Road to 5G Networks: Experience to Date and Future Developments* (OECD Digital Economy Papers 284; OECD Digital Economy Papers, Vol. 284). https://doi.org/10.1787/2f880843-en

Ponce Regalado, F., & Rojas Sifuentes, W. (2010). Promoción y desarrollo de las TIC en América Latina. Proceedings of the 4th ACORN-REDECOM Conference Brasilia, D.F., May 14–15th, 2010.

Portugal-Perez, A., Reyes, J.-D., & Wilson, J. S. (2009). Beyond the Information Technology Agreement: Harmonization of Standards and Trade in Electronics. *Policy Research Working Paper*. https://doi.org/10.1596/1813-9450-4916

Privacy and Cybersecurity Developments in Latin America. (2018). https://www.jonesday.com/en/insights/2018/06/privacy-and-cybersecurity-developments-in-latin-am

Rogers, E. M. (1983). *Diffusion of Innovations* (3rd ed.). Free Press; Collier Macmillan.

Rothschild, E. (1995). What Is Security? *Daedalus, 124*(3), 53–98. https://www.jstor.org/stable/20027310

Saner, R., & Yiu, L. (2003). International Economic Diplomacy: Mutations in Post-Modern Times. *Netherlands Institute of International Relations "Clingendael,"* Discussion Paper No. 84.

Seely, B. E. (2003). Historical Patterns in the Scholarship of Technology Transfer. *Comparative Technology Transfer and Society, 1*(1), 7–48. https://doi.org/10.1353/ctt.2003.0011

Sherman, J. (2022). Russia's War for Control of Global Internet Governance. *SSRN Electronic Journal*. https://doi.org/10.2139/ssrn.4119863

Tekir, G. (2020). Huawei, 5G Network and Digital Geopolitics. *International Journal of Politic and Security, 2*, 113–135.

Thomas, E. (2017, November 13). Taming the 'Wild West': The Role of International Norms in Cyberspace. *E-International Relations*. https://www.e-ir.info/2017/11/13/taming-the-wild-west-the-role-of-international-norms-in-cyberspace/

Vázquez Callo-Müller, M. (2020). *How to Build Interoperability? Conceptualizing the Asia-Latin America Relationship for the Data Economy*. ALADI. https://repositorio.aladi.org/handle/20.500.12909/30826

Voon, T., & Mitchell, A. D. (2019). Australia's Huawei Ban Raises Difficult Questions for the WTO (SSRN Scholarly Paper 3390675). https://doi.org/10.2139/ssrn.3390675

White, C. L. (2015). Exploring the Role of Private-Sector Corporations in Public Diplomacy. *Public Relations Inquiry, 4*(3), 305–321. https://doi.org/10.1177/2046147X15614883

WTO. (1994). *Government Procurement—The Plurilateral Agreement on Government Procurement (GPA)*. World Trade Organization. https://www.wto.org/english/tratop_e/gproc_e/gpa_1994_e.htm

Yoshimatsu, H. (2007). Global Competition and Technology Standards: Japan's Quest for Techno-Regionalism. *Journal of East Asian Studies, 7*, 439–468. doi:10.1017S1598240800002666

Part 2

Power Competition in Latin America

From Television Networks to Outer Space

6 Zapping Channels

Brazil's Impact on the US Dominance in Digital Television in Latin America

In 2006, the US lost part of its technology-related influence in South America. Brazil, one of the Americas' significant economies, decided to update and adapt to Terrestrial Digital Television (TDT) to replace analog television. After a series of negotiations, the Brazilian state introduced the Japanese Digital Television Broadcasting System (ISDB-T) into their territory. It also proposed a series of modifications to establish the ISDB-T*b* locally and pushed for a strategy to make it the predominant system across the continent.[1] Brazil chose the Japanese standard over the expected American ASTC and their long-term partner, Europe, which supported its DVB-T standard, the most disseminated worldwide. The decision sparked a chain reaction, causing other South American countries to follow in accord with the Brazil–Japan strategy.[2]

This is significant because it demonstrates how a middle power, not so often referenced in technology-related discussions, can become a benchmark for Latin America's adaptation. The case was mentioned by public officials and experts interviewed in this book as a situation where neither the US nor China became the leading actor in this significant scenario of technological diffusion.[3] According to the sources mentioned above, its importance lies in the fact that traditional influence cannot explain the results and that they challenged expectations on the diffusion of technology in the region. Conclusions on why this happened vary, but geopolitical and value-related factors seem to have come into play following the three lines of reasoning explored in the previous chapter. In many cases, the wisest choice is to allow systems and technologies to compete, which might lower prices and discriminate the best options for a local environment, but this can also create problems since it might leave individuals marginalized outside of critical technologies that they need to contribute to society or benefit from it.[4] For example, if there were competing and isolated internet systems, this would disrupt people's daily lives and render critical processes useless in the network, such as sending an email if the receiver doesn't use the same technology standards. This is what happens, to some extent, with the PRC's limitation of internet navigation, which doesn't freely allow for communication between its domestic market and the world.

As mentioned in previous chapters, standardization of innovation is a critical process for technology diffusion that involves not only the developers of technology but multiple stakeholders, from the state as a regulator and the companies

DOI: 10.4324/9781003489450-8

that benefit from a state choosing a specific set of qualities to be implemented within their national borders.[5] Nevertheless, standardization contributes to economic, security, and social goals by diminishing expenses, enhancing transparency, and improving market efficiency. The answers to the best standard for a specific technology depend on the actors involved and the economic, political, and sociological system that dominates the environment.[6] During the Cold War, both the West and the Soviet Union relied on technology thought out within their spheres of influence, despite the possibility of superior standards on the other side of the Iron Curtain, and this was part of phenomena such as the Space Race.

Therefore, different states will have different answers to the same question: which standard is better to implement within their borders? In the case of the US, the UK, and France, national sovereignty continues to be linked to the decision-making process. This is known as "mission-oriented", used to describe the connection between a specific technology and achieving goals of national importance, even if this affects the potential reduction of costs and efficiency.[7] On the other hand, "diffusion-oriented states," such as Germany or Switzerland, value the interests of private actors and decentralization and promote human capital from a state position. In these systems, companies are given great power in regularization but are also more vulnerable to competition and shocks.[8]

There are two specific takeaways from this. First, standardization—in any form—is expected to create winners and losers as an externality, since it establishes which companies will hold advantages within a system, either at a national or regional level. Second, the reasoning behind said standardization is not always purely economic. Power and influence—in the form of soft power—can also influence how technology crosses borders.

Since the diffusion process demands not only the strategy of private actors but also the support of the public sector, policy discussions pivot to the domestic and international spheres. In this chapter, I review a significant case: introducing a Japanese standard of digital television in South America,[9] which began with Brazil's domestic innovation policy. I demonstrate that this experience, although not as controversial as the 5G discussion, illustrates the process of technology diffusion and soft power in the Latin American region, the selection of an alternative that escapes the traditional US influence, and the role of Brazil as a regional model for South America, although not entirely a geopolitical leader.

The US (ATSC) and the Introduction of TDT

The predecessor to Digital Television was the *analog system*, and countries that created standards saw commercial and political power in disseminating their system. In the US, the National Television Systems Committee (NTSC) was introduced in 1954 and was adopted by post-war Japan and most Latin American countries. On the other side of the Atlantic, European states also pushed for their standard to globalize. Therefore, the Phase Alternate Line or PAL—a German innovation—was introduced in most of Europe, Asia, Oceania, and Africa. In Latin America, it was preferred by Brazil, Uruguay, and Argentina.

On the other hand, France implemented its system to protect its industry, Sequential Color and Memory (SECAM), which reached the USSR through a deal in 1965, and some parts of Africa.[10] The world was then divided, and companies and providers needed—as a result—to adapt their technologies to the territories they were operating in. Different standards meant inherently different technologies utilized to achieve the same: broadcasting television for a nationwide audience. It is easily arguable that television played a crucial role in society at the time, propagating messages, ideas, and entertainment, and that would be of great importance for political outcomes and policy changes. Therefore, the selection of a standard involved clear ideological conditions that influenced user experience and societies. These technologies were incompatible, so selecting a standard gave certain providers an advantage and closed specific markets toward protectionism, as in France.

As it reached the latter part of the 20th century, the US embarked on a quest to find a renewed television standard for the digital era. This was a process of enormous importance, as the analog system had been in operation since the beginning of public broadcasting, and the world was undergoing significant transformations toward digitalization.

In 1982, The Federal Communications Commission (FCC) enlisted the services of the Grand Alliance (GA), a consortium of electronics and telecommunications corporations, to develop said updated standard. The outcome was the Advanced Television Systems Committee (ATSC), which superseded the National Television Systems Committee (NTSC) in 1993. However, it didn't reflect a unanimous consensus within the GA, a factor that—for many reasons—delayed the final transition and the end of analog television to as late as 2009.[11]

The introduction of Digital Television meant a series of innovative features that improved the experience and the relationship between the user and the broadcaster. For example, the audience could participate in polls, play games, or send and receive emails. On the other hand, the new system allowed for more channels within the spectrum, allowing for a broader range of channels people have access to.[12] In particular, the US system was a good match for their domestic market and environment; it fit businesses' needs and reached rural areas with a low population density.[13]

Nevertheless, significant factors hindered the dissemination of TDT and delayed the transition process. One of those factors was the preponderance of cable television, which reached an essential percentage of the population—more than 60% by 2011—and disincentivized the adaptation to a new system. There was also a lack of coordination and preparation domestically, which was—according to some authors—the result of prioritizing continuation over reform.[14]

Therefore, the US held an advantage as the pioneering actor in the introduction of TDT, but this lack of consistency and the particular demographics the system was created for allowed competing actors to develop alternatives that suited their domestic markets.

The Japanese ISDB-T

As the US introduced its ATSC, other countries began to adapt and create alternatives that better suited their local environments. Japan's answer to digital television was The Integrated Systems Digital Broadcasting (ISDB-T), developed by the Association of Radio Industries and Business (ARIB). This entity was founded only two years after the US changed its system, in 1995, by the Ministry of Posts and Telecommunications, and it gathered an arrangement of institutions, including representatives from the telecommunications, radio broadcasting, and radio equipment sectors, as well as some electricity and gas suppliers, banks, and other types of companies.[15] The committee finally concluded that Japan would need to maintain analog television for as long as the transition process took and that all networks that operated TDT were required to offer high-definition television from the beginning of their transition. This meant operators needed to invest in proper equipment.[16] The government heavily supported the acquisition of TDT modems, the replacement of old televisions, and the diffusion of this technology across the country.[17]

The Japanese standard had several characteristics that met the nation's vision and needs. For example, the ISDB-T features an "Emergency Warning Broadcast System" (EWBS),[18] which is a valuable tool for a country that experiences repetitive natural disasters. It also notifies potential threats, such as missiles from North Korea going over the territory. The ISDB also allows a plethora of small-scale stations to broadcast, intending to expand local reach by creating networks that share atomized audiences.[19] Signals can also reach phones for free, which became attractive for some Latin American countries.

Furthermore, the system is designed to be malleable to the conditions of the environment that adopts it. The standard's advocates amply promote this characteristic. For example, the early warning system was adapted to the Latin American environment as the "EWBS Superimpose Dissemination System," developed in Japan to utilize the ISDB-T technology and existing broadcasting networks.[20] It was highly regarded as countries like Chile often experience natural disasters due to its position in the Pacific Fire Ring.

The European DVB-T

Europe also competed to preserve its dominance in South America, where two significant states, Brazil and Argentina, used PAL for decades. The private telecommunications sector in Brazil was close to Europe since their technology had been adapting to their systems. It was then a suitable and rational thought to believe that the European (DVB-T) could become the preferred system in South America.[21]

Europe began searching for an alternative to analog television in the 1980s. The Multiplexed Analogue Components (MACs) gained traction as a satellite system in the region and to provide high-definition quality. Although the initial project didn't land and ended in 1993 with significant losses for the companies involved, some firms also participated in the process and competed in the US to define the TDT system. The EU Council of Ministers later endorsed a new initiative to achieve digital television in the long run. As a result, they formed the "Digital Video Broadcasting (DVB) Group"—a consortium established in 1993 that consisted of 80 members from the European industry, including public agencies, audiovisual equipment manufacturers, and broadcasters.[22]

The European TDT system was approved in 1997, and a series of actions began, including a patent pool for the DVB standards and an international dissemination campaign. As of 2024, it is the predominant system worldwide, reaching most of Asia, Africa, and Europe. In Latin America, however, it only operates in Colombia.

The Chinese DMB-T/H

China does not participate in the international diffusion competition as the other three systems. The Chinese telecommunications industry is heavily influenced by the Chinese Communist Party (CCP) and state-controlled media, which dominate most channels for information transfer. Television is one of those critically essential channels since it provides unilateral content delivery that can be shaped to serve the state's vision and communicate it to the population. For that and other reasons, in the context of implementing technological standards, China decided to opt for the creation of its own standard rather than adopting a foreign one, as it had previously done with analog television—the DMB-T/H (Digital Multimedia Broadcasting-Terrestrial/Handheld) was its answer.[23] This standard was proposed by experts at Tsinghua University and was compared with two other proposals by the Shanghai Jiao Tong University and the Academy of Broadcasting Science.[24] The initial objective was the domestic diffusion of DMB-T/H, as there were remote areas without access to cable television or other forms of media.[25] In accordance with the experience of former standard creation and political objectives, although the PRC's actions did not conclude with the system's internationalization, it reflects a domestic and international strategy related to technology policies.

The Competition for TDT Diffusion in Latin America

After the US released the ATSC, other states quickly took on the significance of this innovation and invested resources in creating their alternatives. The result was the consolidation of four international TDT standards: the US (ATSC), the European (DVB-T), the Japanese (ISDB-T), and the Chinese (DMB-T). Those states that didn't produce a standard began to evaluate this new technology and the advantages for their local population and industry. Therefore, the models started a race to disseminate their technology further abroad.[26]

A country could choose one of these systems or adapt it to its environment.[27] Still, as it happened with analog television, this decision excluded competing technologies, as Angulo et al. explain:

> These standards are not compatible with each other, and the broadcasts from one system cannot be received by televisions in a different one. This situation has forced countries to choose the standard that is most suited to their

> objectives, and it has also caused standards promoters to compete in order to attract the largest number of countries possible.[28]

Given that the US had already successfully promoted the NTSC standard throughout Latin America, it was reasonable to assume that it would also achieve similar success in promoting its new TDT system. The US had been at the forefront of efforts to promote TDT, making it the obvious choice to lead the charge in spreading the new technology.

Washington aimed to have the ASTC as the dominating system across the Americas. To this end, it began promoting it within regional organizations such as the OAS and the Inter-American Telecommunications Commission (CITEL in Spanish), to which it had privileged access as a member.[29] The US also launched several schemes of investment and funding to approach the region's most prominent actors, especially Brazil and Argentina, which had previously used PAL as its analog system. In 1998, Argentina created a committee to implement digital television, and this committee aimed to deploy the ASTC and start transmitting the same year. The assumption was that the ASTC would provide high-definition quality and that adapting the existing infrastructure and systems at a national level would be possible. The international opinion that came with other countries' experience using the standards was considered trustworthy. As a result, the American proposal was successfully implemented as digital television by the Buenos Aires government.[30]

In October 2003, US and European states announced incentives for Brasilia if the government chose their TDT systems for their domestic market. In the case of Washington, the Overseas Private Investment Corporation (OPIC) promised up to US$150 million in funding if Brazil chose the ASTC system, while Europe also pledged financial support.[31]

The Digital Broadcasting Experts Group (DiBEG) was founded in 1997 in Japan. Organized by broadcasting companies and manufacturers, its objective was to promote the ISDB-T digital television system worldwide.[32] This strategy included training target countries once they selected the system and providing aid and loans.[33] Japan also took advantage of the Emergency Warning Broadcast System, which adapted to Latin America's severe and localized threats and needs.[34]

Brazil originally intended to create the *Sistema Brasileiro de Televisão Digital* (SBDTV), or the Brazilian System for Digital Television in English. This initiative—sometimes referred to as the "nationalist" proposal—was pushed by President Lula Da Silva during his first term (2003–2010) but quickly lost its public and political support, leading to international competitors' involvement in the negotiations.

While the private sector in Brazil openly supported the European alternative, the Government opted for the Japanese one, establishing the ISBD-Tb.[35] The Japanese model was considered more advanced than its competitors and performed better in tests conducted during the evaluation process.[36] Other advantages included the promise of implementing a factory for producing semiconductors, a waiver on royalties, and technical feats such as the implementation of Brazilian-produced middleware[37]—which was later released for free to other partners[38]—the fact that it didn't interfere with private interests as much, and the expression of geopolitical interests.[39]

Even though the decision surprised some experts and businesses in the country, the outcome was timely and was rationally discussed internally in the Brazilian public sector. The broader objective was establishing a regional standard based on Brasilia's decision, supported by Japan, to disseminate ISBD-Tb.

After negotiations, an agreement was reached, and in June 2006, President Lula da Silva signed Decree 5820, which ended analog television and set the framework for digital television. The decree established the characteristics of the Brazilian version of the ISDB-Tb (ISDB or ITU-R Rec. BT 1306 system C), which included:

> The use of the more advanced H.264 standard for digital video coding, as opposed to the MPEG-2 standard used in the ATSC, DVB-T, and ISDB-T standards; The Ginga Middleware, which has been specifically designed for ISDB-Tb (…); [or] the use of the same technology as the one used by the Japanese standard ISDB-T, for coding and modulating digital television signal.[40]

It took some time before Brazil launched its digital TV system, and neighboring countries began considering which model to implement in their territories. The American ATSC had quickly reached Canada, El Salvador, Honduras, Mexico, and South Korea,[41] further confirming the US's constant influence in Central America and the difference between South America and the rest of the continent. For example, Argentina, which implemented the ATSC in 1998, revisited its decision in 2005 when it opened different public discussions that compared the three central competing systems. In 2006, the government created the Commission for the Study and Analysis of Digital Television Systems, which concluded that there weren't significant differences between the offers. At that moment, it seemed that Argentina would lean toward the European alternative. Nevertheless, by August 2009, the Argentinian government declared it would adopt the Japanese system, adapted to the Brazilian reality.[42]

The Argentinian decision has been said to be influential in other regional decision-making processes. By the end of 2009, the European DVB/T was preferred in Europe, Asia, Africa, and Oceania. In South America, it had been adopted in Uruguay and Colombia. However, Uruguay also transitioned to the Japanese system and had adopted the ISDB-T by the beginning of 2011.[43] Others, like Chile and Peru, opened their public tenders and, finally, chose the same Japanese system, considering these competitive advantages and the regional trend. In Peru, the Brazilian-Japanese offer was announced as the preferred option in April 2009 after evaluating all available options.[44] Peru was an essential destination for Japan, as the home of the largest Japanese diaspora—after Brazil and the US—which meant both countries shared a cultural and historical relationship. DiBEG was keen to standardize its disaster prevention system EWBS internationally, so Peru became a milestone,[45] which was immediately part of Japan's public diplomacy and international cooperation strategy.

In the case of Chile, which announced its decision in September 2009, the unique geography of the country and the population distribution were essential factors in the selection of the system.[46] Public officials also claim the importance of soft power and the "beauty contest" system that made this alternative more attractive.[47] Beauty contests refer to evaluating offers not only by their technical characteristics and price but also by factors such as experience, reach, or quality of the project.[48]

Other political actions that reflect the regional conventions on TDT system adoption include a 2009 Mercosur agreement to promote ISDB-T and Argentina's 2010 proposal to create a working group in Unasur to facilitate the diffusion of this technology.[49]

With decision-making processes taking place all across the continent over the years, the ISDB-T system had been adopted by 14 countries in Latin America by 2020. On the one hand, Brazil hoped for a homogenous adoption of the technology in the region, which would help expand its influence but also for economic reasons such as lower prices and efficiency. A public diplomacy operation has directly been promoted from Japan, as the Digital Broadcasting Experts Group or "DiBEG" provided technical support for implementing ISDB-T in those territories.[50] Japan sees South America as a success story, as the European DVB/T reached the most territories during the consolidation phase of this technology.

With the competition and development of these different systems, Latin America became divided into three groups. Those states that followed Brazil and Japan as they persuaded the region to adopt ISDB-T under the argument of economic efficiency or political cooperation; those states that followed the US due to its influence and—in the case of Mexico—the privileged access to the US market; and those who followed the European standard,[51] like Colombia, that values the wide use of DVB/T around the world and implemented its updated version.[52] Countries began their Analog switch-off (ASO) process once they selected and implemented the TDT system. Although some have not completed the process, most states have, up to May 2024, successfully transitioned to a digitalized television.[53]

Latin America's Decision-Making and Rationale

According to the interviews, the existing literature, and official documents and statements issued by the countries that adopted TDT since it first appeared in the late 1990s, more than economic reasoning is needed to explain the development of this technology in the region. Harmonization and industry-related benefits can partially explain Brazil's and Mexico's decisions but not the overall output. Brazil sought to benefit its national companies by adapting the model that better complemented the domestic market. Mexico took advantage of its relationship with the US and NAFTA.[54] As Angulo et al. explain:

> Latin America has had mixed behaviors in harmonization related to economic incentives. For some countries, technical considerations stand before regional preferences or price, whereas others are influenced by externalities, collaboration opportunities, and political relationships with their neighbors.[55]

There is a more technocratic and rational decision-making interpretation of the process. This means that bureaucracy plays a central role in evaluating alternatives, maximizing benefits, and minimizing costs.[56] For some, this explains why Brazil chose the Japanese alternative in the first place. Nevertheless, bureaucracy alone does not explain the diffusion of ISDB-T in the whole region, nor the decision over the lack of solid quality and price differentiation across options.[57]

Regarding economic incentives, some authors have pointed out that it wasn't the economic offer that convinced countries to adopt this system but the industry-related benefits. Brazil wanted to reduce the costs of investments, knowledge acquisition, and technical expenses. At the same time, other countries saw a political and economic advantage in adopting the same standard that their closer trade partners had.[58]

Shared values over what this technology should provide and the effects of both Brazilian and Japanese soft power have also been discussed as compelling reasons for the South American outcome.[59] The use of soft power can be inferred from the emulation of the policy of another state,[60] which is the case for the countries that adopted the ISDB-Tb system because of Brazil's and Japan's diffusion strategies. As a result, since states adopted the system based on Brazilian policy—although Argentinians might argue that they inspired other nations in the region—there is evidence of decision-making processes examining Japan's and Brazil's reputation and perception. As Professor Angela Brandão explains in an article about this topic, the outcome was not a coincidence but a resolution by Brazilian public officials following political motivations. Brazil wanted to put a face to its positive-agenda leadership strategy, which aimed at achieving greater levels of regional integration while promoting a less imperialistic foreign policy. It also aimed to be perceived as a "friendly power" and avoid isolation, as when Brazil chose the PAL analog system. This also allowed it to promote an economic agenda that benefited from said integration. To put this strategy into practice, a series of actions were taken as soon as Brazil implemented ISDB-Tb. For instance, diplomats engaged directly with the countries they were operating in, and delegations that gathered academics, businesspeople, or parliamentary members were sent to neighboring countries to engage with decision-makers or public officials.[61]

Despite Brasilia's relevant role in the TDT standard diffusion process, the nature of Brazil, both as a middle power and a regional leader for South America, is often contested. Brazilians generally believe their country should take a leading role. Nevertheless, they are unwilling to bear the costs required to do so, such as envisioning their neighbors as more than surrounding countries and giving away their independence. Argentina and Mexico also regard themselves as the rightful regional leader, and countries in the middle tend to follow one of those alternatives depending on each scenario.[62] The size and proximity of the state are insufficient to ensure leadership,[63] so more concrete forms of influence gain significant importance. Therefore, Brazil's leadership in transferring the ISDB-Tb system across South America is not the result of long-standing leadership but of a thought-out strategy.

For Japan, Brazil became a partner and a regional actor that could extend its soft power and localize its offer to the needs of local actors. This was important when it came to disseminating the emergency feature of ISDB-T. As the Digital Broadcasting Experts Group (DiBEG) expresses, Brazil became a "key country for the diffusion of unified EWBS throughout Latin America,"[64] as the coordinated testing and Brazil's adoption, as well as efforts toward a unified system, were crucial for other countries' decision-making.[65] Although its original plan did not include deploying a regional diffusion strategy, Japan recognized a strategic advantage in reaching foreign countries with its technology and acted accordingly.

In hindsight—and according to internal sources involved in the process—Brazil wouldn't have achieved its widespread success without the support of a credible partner internationally known for its capability to support advanced technology.[66]

The transfer of ISDB-T from Japan to South America resulted in economic and industrial benefits. Moreover, Brazil's adoption of ISDB-T enabled the country to profit from the technology. However, financial incentives were only one motivation for Japan's interest in this endeavor.[67] The convention seems to be that political motivations carried significant weight in the efforts made by both Tokyo and Brasilia. In the case of Japan, it provided a rare opportunity to enlarge soft power and boost its reputation as a leading telecommunications industry in a region dominated by US influence and increasingly disputed by China.

Conclusions and Implications

The overview of TDT diffusion in South America shows that economic incentives, although mentioned, were not a pivotal reason for countries to choose ISDB-Tb over the American and European alternatives. Most authors agree that there aren't many economic or technical advantages between the TDT systems that compete in the Latin American region. The cost of implementing any of the alternatives, the price of each system's digital modules, and other forms of differentiation that come from price or quality did not appear to be game-changing elements in the technology diffusion process for Latin America.

When comparing this, offerors appeal to other factors that would help them stand out in the competition or resort to their soft power and geopolitical influence. However, relying solely on traditional notions of power fails to recognize Brazil's conflicting status as a regional leader and the unsuccessful US campaign to spread its ATSC system outside Central and North America.

Considering the similarities between the TDT systems and their costs, soft power and value-related factors have been explored as the most likely reasons that compelled South American countries to adopt the Japanese-Brazilian alternative. Japan's status as a renowned technology provider appealed to some states. Additionally, the disaster prevention system incorporated into the alternative was viewed as a valuable public service outside the political or economic factors that could have influenced the decision. Lastly, the notion of a collaborative system among neighboring nations was welcomed by certain countries as a means of bolstering political connections beyond the reach of the US' influence. Therefore, it can be inferred that there is an inter-regional influence when adopting technologies that go beyond tradition and hard power. Countries will not only consider the price of an offer but also will look at what their neighboring governments are evaluating.

This chapter provides an evidence-based example of how regional power and value-related dynamics might shape technology diffusion outcomes in cases involving geopolitical power and tradition. The significance of regional *pioneers* as a model for Latin America's technology transfer decision-making process cannot be overstated. This has also been evaluated in more recent cases of technology transfer, such as the 5G diffusion in the region or the selection of data-managing providers. As we progress through the book, this trend begins to reveal itself as an intriguing phenomenon worthy of further examination.

Notes

1 Brandão (2015), p. 120.
2 Leiva (2010).
3 M. P., personal communication (February 13, 2023); P. G., personal communication (March 8, 2023).
4 Angulo et al. (2011), online.
5 Lopes de Souza and Saboia Lima de Souza (2009).
6 Ergas (1987), p. 192.
7 Ergas (1987), pp. 194–195.
8 Ergas (1987), pp. 209–213.
9 A relevant portion of the background section of this Chapter is based on a 2011 article by Angulo et al., in which the economic reasons to why the Japanese alternative was chosen is clearly outlined.
10 Crane (1978), pp. 273–276.
11 Angulo et al. (2011), online.
12 Alencar et al. (2010), p. 3.
13 Angulo et al. (2011), online.
14 Albornoz et al. (2012), p. 147.
15 de Oliveira and Lessa (2023).
16 Nakamura and Kumabe (2012), p. 235.
17 Nakamura and Kumabe (2012), p. 240.
18 Sakaguchi et al. (2020).
19 Kawai et al. (1996).
20 Sakaguchi et al. (2020).
21 Brandão (2015), p. 129.
22 Angulo et al. (2011), online.
23 Weber (2012), pp. 250–252, 258.
24 Albornoz et al. (2012), p. 31.
25 Weber (2012), p. 260.
26 Lopes de Souza and Saboia Lima de Souza (2009), p. 6.
27 Huang and Liao (2012).
28 Angulo et al. (2011), online.
29 Albornoz et al. (2012), p. 273.
30 Krakowiak et al. (2012), p. 199.
31 Lopes de Souza and Saboia Lima de Souza (2009), p. 8.
32 de Oliveira and Lessa (2023)
33 Brandão (2015), p. 132.
34 Sakaguchi et al. (2020).
35 Brandão (2015), pp. 122–123.
36 Alencar et al. (2010), p. 3.
37 Lopes de Souza and Saboia Lima de Souza (2009), p. 8.
38 Brandão (2015).
39 Krakowiak et al. (2012), p. 197.
40 Alencar et al. (2010), p. 3.
41 Angulo et al. (2011), online.
42 Krakowiak et al. (2012), pp. 201–204.
43 Krakowiak et al. (2012), p. 204.

44 DiBEG (2009), online.
45 Hirose (2016), pp. 7–8.
46 SUBTEL Chile (2009), online.
47 Anonymous, personal communication (March 8, 2023); P., M. (2023, February 13).
48 Paredes (2004), online.
49 Angulo et al. (2011), online.
50 Sakaguchi et al. (2020).
51 Angulo et al. (2011), online.
52 López-Sánchez et al. (2010).
53 *Status of the Transition to Digital Terrestrial Television* (n.d.), by May 1, 2024.
54 Angulo et al. (2011).
55 Angulo et al. (2011), pp. 773–775.
56 Mintz and DeRouen (2010), p. 7.
57 Angulo et al. (2011); Brandão (2015).
58 Lopes de Souza (2009), p. 11; Angulo et al. (2011), online.
59 Brandão (2015).
60 Gallarotti (2011).
61 Brandão (2015).
62 Onuki et al. (2016).
63 Onuki et al. (2016).
64 Sakaguchi et al. (2020).
65 Sakaguchi et al. (2020).
66 Brandão (2015).
67 Yoshimatsu (2007) provides the basis to explain the economic decision-making rationale for Japan's technology dissemination strategy.

References

Albornoz, L., García Leiva, M. T., & Fuertes, M. (Eds.). (2012). *La televisión digital terrestre: Experiencias nacionales y diversidad en Europa, América y Asia* (1o edición). La Crujía Ediciones.

Alencar, M. S., Lopes, W. T. A., & Madeiro, F. (2010). History of Television in Brazil. *2010 Second Region 8 IEEE Conference on the History of Communications*, 1–4. https://ieeexplore.ieee.org/document/5735326

Angulo, J., Calzada, J., & Estruch, A. (2011). Selection of Standards for Digital Television: The Battle for Latin America. *Telecommunications Policy*, *35*(8), 773–787. https://doi.org/10.1016/j.telpol.2011.07.007

Anonymous. (2023, March 8). *Interview with a Former Member of One of Chile's Security Agencies* [In Person].

Brandão, A. S. (2015). Soft Power and Digital Television in South America: The Brazilian Campaign to Promote ISDB-Tb According to Governmental Actors' Narratives. *Intercom: Revista Brasileira de Ciências Da Comunicação*, *38*, 119–138. https://doi.org/10.1590/1809-5844201527

Crane, R. J. (1978). Communication Standards and the Politics of Protectionism: The Case of Colour Television Systems. *Telecommunications Policy*, *2*(4), 267–281. https://doi.org/10.1016/0308-5961(78)90041-1

de Oliveira, H. A., & Lessa, A. C. (2023). Brazil–Japan Relationship: A Partnership? In N. Hamaguchi & D. Ramos (Eds.), *Brazil—Japan Cooperation: From Complementarity to Shared Value* (pp. 21–42). Springer Nature. https://doi.org/10.1007/978-981-19-4029-3_2

Di BEG. (2009, April 24). *Peru Adopts ISDB-T*. Di BEG. https://www.dibeg.org/news/2009/0904peru_adopted/

Ergas, H. (1987). Does Technology Policy Matter. In *Technology and Global Industry: Companies and Nations in the World Economy. Center for European Policy Studies*, (pp. 4–69). National Academies Press.

Gallarotti, G. M. (2011). Soft Power: What It Is, Why It's Important, and the Conditions for Its Effective Use. *Journal of Political Power*, *4*(1), 25–47. https://doi.org/10.1080/2158379X.2011.557886

Hirose, K. (2016). *ISDB-T Activities in Peru*. New Breeze.

Huang, K., & Liao, M. (2012). Analysis of Digital Terrestrial Television Development in Taiwan. *19th ITS Biennial Conference*, Bangkok. https://www.econstor.eu/handle/10419/72538

Kawai, N., Namba, S., & Yamazaki, S. (1996). Performance of Multimedia Broadcasting through ISDB Transmission System. *IEEE Transactions on Broadcasting*, *42*(3), 151–158. https://doi.org/10.1109/11.536574

Krakowiak, F., Mastrini, G., & Becerra, M. (2012). *Argentina: Razones geopolíticas y perspectivas económicas*. La Crujía. https://hdl.handle.net/10016/32992

Leiva, M. T. G. (2010). The Introduction of DTT in Latin America: Politics and Policies. *International Journal of Digital Television*, *1*(3), 327–343. https://doi.org/10.1386/jdtv.1.3.327_1

Lopes de Souza, T., & Saboia Lima de Souza, R. (2009). Building the Digital TV Standard: The Brazilian Experience. *2009 GLOBELICS Conference*. https://citeseerx.ist.psu.edu/document?repid=rep1&type=pdf&doi=f824f80e0862fb26f1372089bf8ec58da71dde32

López-Sánchez, J., Garcia Acero, C. G., Gómez-Barquero, D., & Cardona, N. (2010). Planning a Mobile DVB-T Network for Colombia. *IEEE Latin America Transactions*, *8*(4), 444–453. https://doi.org/10.1109/TLA.2010.5595136

Mintz, A., & DeRouen, K. (2010). Understanding Foreign Policy Decision Making. In *Understanding Foreign Policy: Decision Making*. Cambridge University Press. https://doi.org/10.1017/CBO9780511757761

Nakamura, Y., & Kumabe, N. (2012). *Japón: Servicio universal, alta definición y movilidad*. La Crujía. https://hdl.handle.net/10016/32993

Onuki, J., Mouron, F., & Urdinez, F. (2016). Latin American Perceptions of Regional Identity and Leadership in Comparative Perspective. *Contexto Internacional*, *38*, 433–465. https://doi.org/10.1590/S0102-8529.2016380100012

P., M. (2023, February 13). *Interview with an Inter-American Development Bank Consultant* [In Person].

Paredes, R. D. (2004). Government Concession Contracts in Chile: The Role of Competition in the Bidding Process. *Economic Development and Cultural Change*, *53*(1), 215–234. https://doi.org/10.1086/423259

Sakaguchi, Y., Takahashi, Y., & Sakuma, S. (2020). Activity of Disseminating Japanese EWBS Technology. *Digital Broadcasting Experts Group* (DiBEG), Tokyo, Japan.

Status of the Transition to Digital Terrestrial Television: Countries. (n.d.). ITU. Retrieved May 1, 2024, from https://www.itu.int:443/en/ITU-D/Spectrum-Broadcasting/DSO/Pages/countries.aspx

SUBTEL Chile. (2009, September 14). *Gobierno de Chile adopta norma de televisión digital para el país*. Subsecretaría de Telecomunicaciones de Chile. https://www.subtel.gob.cl/gobierno-de-chile-adopta-norma-de-television-digital-para-el-pais/

Weber, I. (2012). *China: Un entorno multiplataforma entre el Estado y el mercado*. La Crujía. https://hdl.handle.net/10016/32993

Yoshimatsu, H. (2007). Global Competition and Technology Standards: Japan's Quest for Techno-Regionalism. Journal of East Asian Studies, 7, 439–468. https://doi.org/10.1017/S1598240800002587

7 Latin America's Private Data Market and Tech Sovereignty amid Geopolitical Power Competition

The transit, management, protection, and ownership of private data are modern issues and priorities for many institutions. There are initiatives in civil society, the private sector, and states to answer to existing needs under frameworks that range from the infrastructure that makes the transfer of data possible to digital environments in which information flows and is utilized. All these actions are taken by assuming that private data is *strategic*,[1] but the nature of who owns said information varies. In some systems, it is believed that the entities that gather the information are entitled to its use, exploitation, and profit. This trend has been challenged by new regulations that entitle information to individuals and closely limit the scope of private data use.

Consequently, a second framework for information ownership relates to the idea that individuals own their data, and this is just shared under a stipulated agreement in the form of a contract or clear terms of service. The European GDPR establishes strict limitations to the management of personal data, and it should be "processed lawfully, fairly, and in a transparent manner in relation to the data subject."[2] This model has since been a referent for other states looking forward to updating their privacy regulations or seeking the economic benefits of having a homologated regulation with Europe.[3]

A third framework related to security and sovereignty perceptions can be interpreted as the assumption that states have control over data supported by legislation or effective control mechanisms. This is the case for China, as reflected in the 2017 Cybersecurity Law, which puts sovereignty at the forefront of the national strategy and requires users to align with their national strategy, both within and outside the mainland.[4] In Article 6, the law states that:

> The State advocates sincere, honest, healthy and civilized online conduct; it promotes the dissemination of core socialist values, adopts measures to raise the entire society's awareness and level of cybersecurity (...).[5]

At the same time, Chinese regulations require citizens to follow certain behaviors and content rules and compel companies and government organizations to monitor content concerning Chinese citizens. The law also affects telecommunication companies and service providers,[6] effectively putting national interests over the

DOI: 10.4324/9781003489450-9

economic activities of these firms. Critics and analysts often refer to this argument when they warn about potential intervention and security issues related to PRC investments in technology.[7]

Discussing private data and how companies and states treat citizens' entitlement to privacy is a long-standing debate in Latin America. The wide range of democratization levels, development, access to the internet, and media literacy mean that generalization is difficult, but certain phenomena spread across the region. The main question involved with the transfer of information is related to how a state assesses factors such as security, private data, and economic benefits. In other words, the decisions made by a state should consider geopolitics, values, and the price of the technology that will be deployed within their territories. An essential concept in the definition of privacy and data for Latin America is the writ of *habeas data*, from the Latin "[we command] you have the data," a constitutional right that protects citizens from the incorrect collection and use of their data and allows citizens from some countries to challenge their governments and request changes in the information they hold.[8] According to the Organization of American States' Data Protection Documentation:

> Habeas Data is an action that is brought before the courts to allow the protection of the individual's image, privacy, honour, self-determination of information, and freedom of information of a person. This is a mechanism of action that gives the aggrieved person to file a complaint with the judicial system to stop the abuse of personal data.[9]

The debate about data is unequivocally linked with data infrastructure and the hardware needed to facilitate data storage and management. This has permeated influence-building debates. Technology diffusion trends show increased states' importance to international politics and how policies mirror the changing political theater. In 2012, Australia banned Chinese providers from participating in telecommunications systems, marking a line between technical and economic assessment and geopolitics.[10] In 2018, the same state provided funding to connect with the Solomon Islands and Papua New Guinea, limiting Huawei's expectations to build a similar project. Washington has been actively involved in diplomatic efforts to restrict Chinese technology infrastructure in the Pacific.[11]

Following the assumptions and theoretical framework developed in previous chapters, I explore several of these cases, including the Asia-to-South America *transpacific cable* in which Chinese and Japanese offers were considered; the discussion over public citizens' data, and the implementation of passports in Chile, again decided between a Chinese and a European company; the implementation of identification technology in Venezuela; and the growing cloud and internet traffic markets, in which the US has had an advantage. These cases illustrate how technologies have spread throughout Latin America in recent years, highlighting the impact of different governmental, geopolitical, and economic alliances. I also refer to connectivity problems in the region, as mentioned by experts and public officials I talked to during the data-gathering phase of this book. The analysis shows how domestic and international affairs are interconnected in an environment where private and public organizations advocate for specific interests. Each case provides insight into the primary factors decision-makers consider when selecting a particular provider.

The Discussion behind the Asia-to-South America Submarine Cable

Submarine cables are deployed all around the globe, transferring information and enabling communications at an outstanding speed. The history of submarine cables dates to the mid-1800s when they were used to facilitate the extension of telephone and telegraph lines.[12] Telecommunications are becoming a pivotal part of development and access to services. They are also a strategic point of contingency within a geopolitical context in which states are looking to secure their information-transit channels.[13] For states, securing through physical infrastructure related to the safe transit of information and protecting intangible private data has become a security, legal, and normative debate.[14] The hardware and software employed in the effective development of cybersecurity are blurry and difficult to categorize, as both the infrastructure and data are critical when analyzed from an international relations perspective,[15] and submarine cables connect hubs under identifiable influence. Coherently, a line has begun forming between networks connected to China and those the US and its allies attempt to keep separated from Beijing's interference.[16]

A quick view of these cables' cluster points and trajectory reveals that Latin America is significantly less interconnected than other developing regions.[17] This means that there is an open opportunity for new projects to emerge if they can overcome the limitations of geography and population distribution. Connecting the Asia-Pacific region brings about these challenges, as any project requires deploying a cable through a wide area to connect countries that might not provide high returns. However, in the case of South America, Chile is a strong candidate for this connectivity initiative. Chile has the longest coastline in the continent, is one of the countries with the fastest internet in the world, and is widely recognized for being a bridge between Latin America and the Asia Pacific, as a pioneer state in the establishment of bilateral economic agreements and a "broker"—filling a strategic and diplomatic vacuum—for both regions,[18] but especially for China. This made it an excellent candidate to consider a cable to connect Asia with South America, as well as a case study of the effects of geopolitics in technology sharing across states.

Considering all these conditions, several proposals were supposed to cross the Pacific Ocean from South America to Asia. Becoming a connectivity hub for South America was an attractive possibility for the Chilean government. Therefore, the state was open to receiving offers to build infrastructure that could help support those objectives.[19] Huawei had already built part of the country's optic fiber system in the south of Chile[20] along with the telecommunications provider WOM, and it was seeking to expand its influence and market in the region. Company representatives and diplomatic figures approached the Chilean Telecommunications Sub-Ministry (Subtel) with a plan to build a cable to connect China and Chile.[21] Both countries enjoy stable diplomatic and economic relations, so the project was an excellent opportunity for both sides to follow their objectives. However, the deal that Huawei and Beijing were trying to secure—through a direct deal discussed by authorities—involved PRC investment and demanded that a Chinese company build the final cable. This would give them a preferential position in a project without considering any other alternative routes or an open bidding process.[22] The restriction did not land well on the Chilean side, and conversations were halted despite the Huawei feasibility document being of good technical quality.[23]

Consequently, the Development Bank of Latin America: Corporación Andina de Fomento (CAF), co-financed a project to see whether the submarine cable was feasible,[24] and the Chilean government began to receive offers that considered economic, geographical, and technical conditions to implement a cable connecting Asia and South America.[25] Several options were obtained, and Huawei presented an offer that seemed the most economically savvy at the time.[26] The Japanese company NEC, which had also provided a pre-feasibility report, competed by proposing a cable that did not begin in Shanghai but connected Japan with Australia and then Australia with Chile, effectively replacing the Chinese basepoint as an option. The Japan Bank for International Cooperation and the Japan ICT Fund (JICT Fund) were expected to finance part of the US$400 million needed to complete the project.[27]

The Chilean decision to choose Japan over China was met with surprise,[28] considering the economic capabilities and influence that the PRC has over the South American state. The results were interpreted as a financial and geopolitical decision to avoid the risks and political consequences of engaging with Chinese technology.[29] Since Australia is trying to limit Chinese presence in its telecommunications system, it would be sensible to assume that Beijing would be left out of possible future contracts. However, this is rejected internally by some public officials within the Chilean ministry who instead claim that Beijing's pressure to only go with Chinese companies, as well as the need for open and more transparent competition.[30]

Later developments put US-based companies at the forefront of the project. In January 2024, Google announced it would be a structural partner in constructing the Humboldt submarine cable, connecting Chile with French Polynesia.[31] The company will create a Joint Venture with the Chilean development fund Desarrollo País, and the project is expected to be completed by 2026.[32] Moreover, it has been discussed that Subcom, another American company, will operate as the system supplier.[33] This was openly celebrated by the US Spokesperson's Office, highlighting the contribution that the government would make to the endeavor—US$15 million by the Bureau of Cyberspace and Digital Policy (CDP)[34]—and further confirming the geopolitical impact and significance of the news for North American power.

There are two critical takeaways from analyzing this process. On the one hand, the sometimes-overlooked role of middle powers with solid diplomatic ties and reputation—as in the case of Japan. On the other hand, the US' underlying power in terms of technology companies' reputation and possible—although sometimes denied—direct geopolitical influence. I argue that the outcome of the Humboldt submarine cable was likely the result of a combination of factors. In Chile, institutionality is comparatively more robust than in other countries in Latin America, but this does not prevent it from engaging in geopolitical quarrels, and this is often a source of internal debate due to the slow communication between Foreign Affairs or intelligence agencies, and the technical institutions in charge of assessing offers trying to uphold neutrality in their decision.[35]

The Chilean Passports Crisis amid the Geopolitical Tensions between the US and China

Even if the discussion over Chinese tech companies in Latin America was not as controversial as in other regions, there are quantifiable examples of states being forced to make decisions based on value, political, and security-related grounds. That was the case of the public tender in which the Chilean Identification and Citizens Registry (Registro Civil) looked for companies that could provide the passport and national identification system management and manufacturing. The competing providers were Aisino, a Chinese State-connected company, Veridos, Telefónica Chile, Sonda-Thales, a Chilean and French venture, and IDEMIA, a French company heavily involved in information and recognition technologies in the region—also the provider of Chilean passports for more than a decade. The Chinese alternative won with the most economically convenient offer,[36] but some government departments, intelligence agencies, Civil Society organizations, and the private sector immediately challenged the decision.

The first reason to challenge the result was security concerns over the country's citizens' data. This was particularly relevant since the public tender came out during a time in which the debate over Chinese technology was also present in other contests, such as the 5G spectrum auction, and during which concerns over Chinese influence in Chilean technology were increasing.[37] Among some public-sector officials, skepticism over China as an investor was also growing, although these concerns were mostly unofficial conversations and were not reflected in the state's policy.[38] This mirrors an overall awareness that is sectorial to those sectors of the government involved in the international discussion but is not always shared at a technical or procedural level.

The second reason for concern was Chile's previous agreements to allow citizens to access visa waivers to enter the US.[39] The FBI had already created a document detailing the Chinese company claiming they were related to "backdoor malware"[40] or the incorporation of covert code capable of accessing information without proper identification. Consequently, upon hearing the public tender results, diplomats and US government representatives raised private and public concerns over how this decision interfered with their state's security objectives.[41]

Finally, the decision showed a lack of communication between the Registry and the Foreign Ministry—which was holding conversations with the US and opposed the Chilean Identification and Citizens Registry decision under the complaint of lack of inter-communication within the government apparatus.[42] Despite Chile's reputation as a well-institutionalized country, coordination between different agencies can be affected by opposing interests or because sometimes there aren't direct channels to share relevant information in time. In this case, the Civil Registry aimed to reduce production costs and inefficiencies, while security agencies were concerned about potential geopolitical and domestic risks posed by the Chinese alternative.[43]

These factors resulted in a debate on how the decision would affect Chile's diplomatic relations with the US. However, there were regulatory and legal reasons for canceling the Aisino contract. The company had invested significant resources into lobbying key actors,[44] but it was also accused of providing false information and missing mandatory documents in its proposal. According to a petition for the dismissal and annulation of the process interposed by Thales-Sonda, the Chinese proposal also had insufficient quality assurance certificates. It also provided a technical offer that did not meet the required standards.[45] Finally, there was enough evidence to claim that Aisino had previously implemented several software systems with backdoors that were effective cybersecurity threats.[46]

The annulation of the awarded public tender brought a debate on the impact of the broader geopolitical rivalry between the US and the PRC in Latin America. Several experts accused "political neglect" and "clumsiness" during the process due to a lack of coordination and geopolitical perspective before communicating the public tender results.[47] Aisino fought back and threatened to take legal action against the Chilean government. At the same time, the Chinese embassy acted in favor of the company and made internal calls to different government agencies, hoping to pressure and change this decision.[48] The Ambassador also complained publicly that the decision went against Chilean non-discrimination principles and could affect the relationship between the PRC and the South American nation.[49] Despite complaints, the Chinese company was fined 100 million Chilean pesos by the Civil Registry for failing to comply with bidding terms and violating the integrity pact required for participating in the Chilean public tender market.[50]

There are two interpretations of this case's outcomes. Some view it as necessary due to the threat posed by a company with a history of engaging in illegal data-related activities. The company also presented a flawed offer with missing documentation. For others, it was a geopolitical and rational decision to follow the US' influence to maintain the exclusive and rare benefit of being exempt from a visa to enter the US.

"Carnet de la Patria" and Venezuelan Identification Contract with ZTE

Unlike the case of Chile, which exhibits strong diplomatic and economic relations with the US and the PRC, other nations have a much more apparent preference due to political and ideological reasons. Venezuela—a country often considered among the poorest and least democratic of the region—has relied on Chinese investments and support in many areas, including citizen control. El *Carnet de la Patria,* or the *Homeland ID-Card,* is a system manufactured by ZTE, a Chinese company, and used by the government to follow its citizens' behavior.[51] The Chavez regime was interested in China's identification program from 2008 when a delegation traveled to Shenzhen to learn about the interaction between technology and citizen control. The idea immediately caught the Chavez government's attention, and despite some internal concerns—according to a notable Reuters piece on the matter—direct negotiations led to an implementation plan to adapt Beijing's vision to the South American country.[52] It took about a decade to concrete, but in 2017, the Venezuelan administration hired the Chinese firm ZTE—widely known for its telecommunication services—to create a national identification system under *Carnet de la Patria*. For Venezuelans, having the card became almost necessary, as it is required to access social assistance, such as food provisions under the CLAP packages, government benefits, and oil subsidies. By 2022, around 18 million citizens had gotten the card,[53] many due to pressure or need. The system has developed since its launch and is now obligatory to vote. It was required to get a vaccine during COVID-19 and a digital wallet.[54] The *Carnet de la Patria* is considered a technological tool that the Nicolás Maduro government has used to remain in power despite economic and political collapse. It also served as a cautionary tale for other countries, not on how China could use the data but on how political alliances could be exploited to perpetuate authoritarianism in the region. Since ZTE didn't need to compete with any other company and since the negotiations were conducted directly, it is safe to assume that geopolitical influence was relevant, if not central, in the selection and implementation of this program. This has also impacted the reputation that the company has garnered in other contexts in the region. Unlike Huawei, ZTE has had a mild penetration in the public and private markets.

The Private Data and Storage Market and Cybersecurity in Latin America

There are cases in which private data-related companies directly invest without the objective of securing a nationwide system or a government contract. *Data centers*, or physical locations that store computer machines that process and keep data safe, have become indispensable for states, businesses, and clouding services to allow for collaborative work and backup of critical information. Many user-level services, from emailing to banking, require data to be stored securely. In other words, data centers are part of day-to-day user interaction with the internet, allowing digitalization to be possible nationwide. However, cybersecurity against digital threats is only sometimes guaranteed by these service providers or the public system, and this creates direct problems for states, as attacks on public sector databases can cause significant losses.[55] This became a pivotal issue during the COVID-19 pandemic (2020–2022) since most states in Latin America were pushed to digitalize services, enhance their capabilities, and make use of their internet channels—in some cases, without proper preparation and infrastructure.[56] In 2020, Latin America had over 700,000 malware attacks daily. The three most affected countries were Brazil (56.25% of all attacks), México (22.81%), and Colombia (10.20%). About 66% were related to private entities, and the rest were attributed to other criminal activities and attacks on a governmental level.[57]

Considering the increasing importance of storing information and enabling digital services, as well as concerns over cybersecurity threats, the responsibility invested in the private sector in affairs related to individuals' data also increases. The role of the private sector once again reflects the discussion in Chapters 3 and 4 regarding its involvement in a state's technology-related policy[58] and its interaction with the state's foreign policy—as well as Robert Putman's two-level games.[59]

Data centers are primarily private, and the origin of their parent companies varies. Several mapping services attempt to mirror said centers' locations and other essential details.[60] I found certain inconsistencies between these platforms related to the specific number of locations—with some missing relevant projects. Still, they are helpful in getting a general picture of how these data centers are deployed in Latin America and what the companies involved in the data management market in the region. Among the most well-known ones is Google, which opened a data center in Chile in 2012 and has one in Uruguay as well—the two relatively stable countries of the South American region. American companies like Microsoft, Amazon, and Equinix are also interested in installing regional centers.

In contrast, the PRC through Huawei, centers, is involved in the data center market, with its locations in Mexico, Buenos Aires, and Sao Paulo. It also shows a presence in Lima and Santiago de Chile.[61] Some telecommunication companies, such as Claro and several Brazilian firms, also have a significant presence. According to experts, although the presence and operation of data centers are not often included in geopolitical and security analyses, the potential impact of misusing that information—and data sovereignty—is concerning for both a state and individuals.[62] Accountability over the transit and use of information is managed within each state. Still, many follow specific standards, such as the OAS cybersecurity framework and the concept of habeas data, or mirror particular characteristics of the European GDPR—that they have embedded to harmonize their system with the European market. I argue that the competition over data transit and data centers in Latin America has not been as controversial as 5G or the transpacific cable is likely to be politicized in the future due to the nature of the competitors and the notion of private data and data-related services as critical for states' notion of security.

Other Information Technology (IT) Trends in Latin America

Along with the location and origin of data centers, analysts point out that Latin America has a connectivity problem and is heavily reliant on the US.[63] Over 83% of Latin America's internet capacity goes through the US.[64] This vulnerability has been identified and is gradually addressed through investments and state initiatives.[65] Argentina and Uruguay are intensifying their links with Brazil over the US, turning Brazil into a central interconnection hub for South America—itself connected to the US.[66] Coherently, Chile has emerged as South America's door to Asia through the Pacific Ocean.

Development might be affected without inter-regional connectivity and solid and secure infrastructure, especially after the COVID-19 pandemic increased digitalization challenges.[67] Improving regional integration could benefit regional hubs, attract investment, and generate local content based on specific digital skills.

States in the region are likely to follow the behavior of their neighbors to harmonize their technology and benefit from the potential development that comes with accessing fast and safe networks. This is something that the Colombian government has openly tried to achieve in recent years.[67] For the Bogotá government, cybersecurity has become a priority, considering that it experiences a disproportionate number of cyberattacks compared to its population size.[68] To better screen potential providers of digital and technology services and try to prevent contracts with providers that do not ensure the safe protection of data, the *Ministerio de Tecnologías de la Información y las Comunicaciones* (MINTIC) or Ministry of Communication and Information Technology has developed a protocol aim at screening potential partner companies during the bidding process. Colombian elite members and decision-makers also follow Washington's assessment of potentially dangerous technology firms. These warnings may influence their opinion but may not always be reflected in state policies.[69]

Latin American countries are increasing their efforts to strengthen their cybersecurity laws. Argentina, Mexico, Venezuela, and Uruguay have updated their frameworks or created specific institutions to sustain and respond to digital threats. However, the responsibility lies in different government authorities depending on the state. Moreover, organizations like the Inter-American Development Bank actively support and fund initiatives to implement effective cybersecurity frameworks. However, there is still a lot of work to be done to increase awareness and build a solid cybersecurity system at a regional level.[70]

Both the US and the PRC have expressed interest in assisting in designing the legal and technical frameworks under which Latin American states define and apply cybersecurity and data-protection-related policies. A pivotal case to illustrate this is Argentina. For instance, in 2018, Huawei sponsored Argentinian high-level officials to attend the Smart City World Congress in Spain. These individuals were Karina Graciela Giusti, Argentina's Director for the initiative "País Digital,"[71] and Andrea Carolina Pazos, the Innovation, Research and Control director of that same initiative.[72] In that same year, the Zhejiang province financed a trip for the General Coordinator of Technical Matters at the National Communications Entity (Enacom) to assist the Fifth World Internet Conference,[73] a Chinese initiative that aims at spreading the Chinese Communist Party vision of nation-governed internet across the world.[74]

Washington has also paid for Argentinian public official trips and invited them to US-based training programs to promote their vision of the potential risks of information technology and how regulation should consider these elements. In 2019, Enacom's Radio-electric planning and management chief was invited to one such training program titled: "5G—The Path to the Next Generation; Digital Transformation: Unlocking the Potential of IoT; Regulatory Principles and Best Practices; Creating a Regulatory Environment for Cloud Services; Enabling the Full Value of Wireless Connectivity."[75] Like many other states in the region, Argentina must balance its diplomatic relations with both powers. The nature and political affiliation of the governing party affect the level of engagement, which has caused tensions in several opportunities. In the next chapter, I dive into the case of the Chinese deep space observation center built in Neuquén, Patagonia, by a PRC institution that is directly linked with the People's Liberation Army (PLA) and how this situation increased the tension and pressure from both Beijing and Washington.

Conclusions and the General Data Protection Scenario

In this chapter, I examined the evolving landscape of data protection and technology sovereignty in Latin America, particularly under the influence of geopolitical powerhouses such as China and the US. To this end, I presented critical cases that reflect the region's strategic importance in the global digital economy and the implications for international relations. The Chilean transpacific cable, in which Huawei competed with NEC (Japan) and later became a joint effort with Google, is a critical example of the competition for influence in the region, confirmed by sources in the private and public sectors interviewed for this book. It also helps illustrate Chile's political priorities and difficulties in navigating and balancing a good relationship with the US and the PRC. This was further proven when reviewing the case of Aisino and the Chilean passport controversy, which, on the surface, mirrors the direct competition between Chinese, European, and Latin American companies. Still, it involved an inter-institutional lack of communication and coordination and possible negligence in reviewing an incomplete offer. Some also point to the US' influence over the final decision, emphasizing the geopolitical considerations.[76]

Regarding geopolitical influence, shared values, and economic incentives, the Venezuelan ID-card contract with ZTE helps frame a different scenario in which the state has an apparent ally and supporter. Direct negotiations and the government's explicit interest in creating a control system that mirrors the one China is implementing make this project a sample of how citizens can be subjected to surveillance through technology.[77]

This chapter also discusses the strategic importance of data centers, cybersecurity, and the frameworks governing data protection. The concept of habeas data, a principle in Latin American countries to protect citizens' data, is explored to highlight that the region has a normative and value-based framework to understand the use of personal information. Finally, I will explore the US' presence in the private cloud sector, as American companies such as Google, Microsoft, and Amazon have established data centers in Latin America. However, it is essential to acknowledge that Brazil has increasingly become a digital and connectivity hub for South America, as several cloud and data services are based there and reach other vital countries such as Uruguay, Argentina, and Chile—which is also aiming at gaining importance in this area.

The cases presented in this chapter explore several critical implications for international relations research. The involvement of China and the US in Latin America's technology and data infrastructure demonstrates how technological advancements are becoming new arenas for geopolitical competition. This reflects a shift from traditional power struggles to more nuanced battles over digital and cyber sovereignty. In terms of sovereignty and control, I attempted to illustrate different approaches to data sovereignty. China's model emphasizes state control and aligns technological infrastructure with national strategies, while the US model tends to support private sector dominance with oversight mechanisms. These differing approaches influence how Latin American countries navigate technological dependencies and alliances. Overall, states have made an active effort to harmonize their regional systems and follow the European GDPR. Adopting GDPR-like regulations and implementing cybersecurity frameworks reflect broader trends in international norms and standards. These frameworks influence how states interact with each other and multinational corporations, shaping the global data protection landscape.

In terms of regional coordination and security, I identify a consistent narrative across experts, public officials, and private sector members on the lack of regional coordination in Latin America regarding cybersecurity and data protection. This gap presents both a challenge and an opportunity for international collaboration and policy harmonization, which are crucial for enhancing regional security and resilience against cyber threats. Despite the relatively high internet penetration in Latin America, specifically in South America, unresolved challenges exist in ensuring stable digital connectivity. Additionally, there is a lack of coordination between states to create efficient networks and hubs. Building telecommunications infrastructure, updating data privacy policies, and ensuring the secure traffic of personal data in an environment where private companies hold a central role presents significant challenges.

Moreover, geopolitical tensions and debates over the state's role in safeguarding these networks further compound the complexity for stakeholders and end-users, who are consistently impacted by these decisions. The COVID-19 pandemic is often mentioned as a pivotal point in transforming digital services in Latin America, forcing states and citizens to change their behavior. In some cases, they were unprepared regarding cybersecurity, internet reach, and capacity.

Regarding economic and strategic interests, decisions are considered the most technical and efficient alternative for technological infrastructure and data protection. Still, outcomes cannot only be explained by those factors. Countries must balance the benefits of modernizing their digital infrastructure with the risks of becoming overly dependent on foreign powers and maintaining good relations with the US and the PRC. This task has become more challenging due to the escalating tensions between those states.

In that context, I propose that the explored cases prove that China and the US play pivotal roles in shaping the current Latin American data protection scenario, each promoting distinct models and strategies. Latin American countries are at a crossroads, needing to balance the technological benefits offered by both China and the US against the risks associated with data sovereignty and geopolitical dependencies.

As digital sovereignty becomes more critical, Latin American countries may increasingly seek to develop their technological capabilities and reduce dependency on external powers. This could involve fostering local tech industries, investing in regional data centers, and implementing comprehensive cybersecurity strategies. Continued engagement with international organizations and adherence to global data protection standards can help Latin American countries navigate the complexities of digital geopolitics. Collaborating with diverse partners, including European nations and middle powers such as Japan, can provide alternative pathways for technological development and data protection.

In conclusion, China's and the US' strategic maneuvers significantly shape the evolving data protection scenario in Latin America. Understanding the implications of this geopolitical competition and fostering regional and international cooperation will be crucial for Latin American countries to safeguard their digital sovereignty and achieve sustainable technological development.

Notes

1 J. S., personal communication (June 3, 2024).
2 *General Data Protection Regulation (GDPR)—Legal Text* (2016).
3 Ryngaert and Taylor (2020).
4 Creemers et al. (2017).
5 Creemers et al. (2017).
6 Creemers et al. (2017).
7 J. S., personal communication (June 3, 2024); P. Z., personal communication (June 14, 2024).
8 Lode (2019), p. 43.
9 OAS (n.d.).
10 Ragan (2012).
11 McGeachy (2022).
12 McGeachy (2022).
13 Serentschy (2024), Chapter 4, online.
14 McGeachy (2022).
15 Carr and Lesniewska (2020).
16 Ting-Fang et al. (2024), online.
17 *Submarine Cable Map* (n.d.).
18 Schulz and Rojas de Galarreta (2020).
19 International Finance Business Desk (2020); Biblioteca del Congreso Nacional de Chile (2020)
20 Diario Financiero (2018).
21 D. G. and A. O., personal communication (March 8, 2023).
22 D. G. and A. O., personal communication (March 8, 2023).
23 T. P., personal communication (February 27, 2023).
24 International Finance Business Desk (2020).
25 SUBTEL Chile (2019).
26 O' Ryan (2018).
27 Submarine Networks (2024).
28 Hirose and Toyama (2020).
29 McGeachy (2022); Miller, J. B. (2020); More Subsea Cables Bypass China as Sino-U.S. Tensions Grow. (2024).
30 D. G. and A. O., personal communication (March 8, 2023).
31 Submarine Networks (2024); Chilean Government Partners with Google to Build Humboldt Subsea Cable. (n.d.).
32 BNamericas (2024).
33 Submarine Networks (2024).
34 Office of the United States Spokesperson (2024).
35 T. P., personal communication (February 27, 2023).
36 Cárdenas and Fajardo (2021).
37 Anonymous, personal communication (March 8, 2023).
38 T. P., personal communication (February 27, 2023).
39 Artaza and Nogales (2021).

40 *Accion de Impugnación. Unión Temporal de Proveedores SONDA THALES contra Servicio de Registro Civil e Identificación* (2021), p. 54.
41 Ex-Ante (2021).
42 T. P., personal communication (February 27, 2023).
43 Anonymous, personal communication (March 8, 2023).
44 *De Bernardo Caro, de Lota Protein, a Jorge Burgos* (2021).
45 *Accion de Impugnación. Unión Temporal de Proveedores SONDA THALES contra Servicio de Registro Civil e Identificación* (2021), p. 5.
46 *Accion de Impugnación. Unión Temporal de Proveedores SONDA THALES contra Servicio de Registro Civil e Identificación* (2021), pp. 54–55.
47 Carrasco and Toledo (2021).
48 T. P., personal communication (February 27, 2023).
49 Saravia and Arteaga (2021).
50 Saravia (2022).
51 Corrales (2020), p. 43.
52 Berwick (2018).
53 Corrales (2020), p. 43.
54 Ellis (2022), p. 27.
55 Flor-Unda et al. (2023).
56 Park et al. (2022), pp. 10–12.
57 Flor-Unda et al. (2023).
58 Eriksson and Giacomello (2007).
59 Putnam (1988).
60 *Global Data Center Map* (n.d.); *Datacenters World Map* (n.d.); *Data Center Map—Colocation, Cloud and Connectivity* (n.d.).
61 *Huawei Cloud Regions: Role, Positioning, and Division Strategy* (n.d.).
62 M. P., personal communication (February 13, 2023), A. G. Z., personal communication (March 13, 2023).
63 A. G. Z., personal communication (March 13, 2023).
64 Gayo (2023), online.
65 Echevarría (2020).
66 Gayo (2023), online.
67 S. S. et al., personal communication (May 5, 2023).
68 Flor-Unda et al. (2023).
69 R. M. L., personal communication (March 16, 2023).
70 Flor-Unda et al. (2023).
71 Government of Argentina (2018: 1).
72 Government of Argentina (2018: 2).
73 Government of Argentina (2018: 3).
74 WIC (2022), online.
75 Government of Argentina (2019).
76 T. P., personal communication (February 27, 2023); D. G. and A. O., personal communication (March 8, 2023); McGeachy (2022).
77 Berwick (2018).

Bibliography

Accion de Impugnación. *Unión Temporal de Proveedores SONDA THALES contra Servicio de Registro Civil e Identificación* (Tribunal de Contratación pública 2021). https://www.diarioconstitucional.cl/wp-content/uploads/2021/08/Demanda-Thales.pdf

Anonymous. (2023, March 8). *Interview with a Former Member of One of Chile's Security Agencies* [In Person].

Artaza, F., & Nogales, D. (2021, November 16). *Gobierno anula licitación de pasaportes a empresa china Aisino tras advertencia de Cancillería y EE.UU.* La Tercera. https://

www.latercera.com/pulso/noticia/pasaportes-el-registro-civil-deja-sin-efecto-la-adjudicacion-de-aisino/OZQEJBQUZ5CHVHNAMSJWLXSK5Q/

Berwick, A. (2018, November 14). *A New Venezuelan ID, Created with China's ZTE, Tracks Citizen Behavior*. Reuters. https://www.reuters.com/investigates/special-report/venezuela-zte/

Biblioteca del Congreso Nacional de Chile. (2020). *El despliegue del cable transoceánico que conectaría a Chile con el Asia a través de Australia—Programa Asia Pacífico* [Text]. Observatorio Asiapacifico; Biblioteca del Congreso Nacional de Chile. https://www.bcn.cl/observatorio/asiapacifico/noticias/despliegue-cable-transoceanico-chile-australia

BNamericas. (2024). *The Next Steps for Chile's Humboldt Cable*. BNamericas. https://www.bnamericas.com/en/interviews/the-next-steps-for-chiles-humboldt-cable

Cárdenas, R., & Fajardo, D. (2021, July 30). *Aisino: La empresa estatal china que lidera la licitación de las nuevas cédulas de identidad y pasaportes chilenos*. La Tercera. https://www.latercera.com/earlyaccess/noticia/aisino-la-empresa-estatal-china-que-lidera-la-licitacion-de-las-nuevas-cedulas-de-identidad-y-pasaportes-chilenos/FXZEKOATWVEMHFSIN2BDG3UX5U/

Carr, M., & Lesniewska, F. (2020). Internet of Things, Cybersecurity and Governing Wicked Problems: Learning from Climate Change Governance. *International Relations*, *34*(3), 391–412. https://doi.org/10.1177/0047117820948247

Carrasco, R., & Toledo, M. (2021). *Expertos critican "descuido político" y advierten que se marca un precedente*. Diario Financiero. https://www.df.cl/economia-y-politica/pais/expertos-critican-descuido-politico-y-advierten-que-se-marca-un

Chilean Government Partners with Google to Build Humboldt Subsea Cable. (n.d.). Submarine Networks. Retrieved June 24, 2024, from https://www.submarinenetworks.com/en/systems/trans-pacific/humboldt-cable/chilean-government-partners-with-google-to-build-humboldt-subsea-cable

Corrales, J. (2020). Authoritarian Survival: Why Maduro Hasn't Fallen. *Journal of Democracy*, *31*(3), 39–53. https://doi.org/10.1353/jod.2020.0044

Creemers, R., Webster, G., & Triolo, P. (2017). *Translation: Cybersecurity Law of the People's Republic of China* (Effective June 1, 2017). DigiChina. https://digichina.stanford.edu/work/translation-cybersecurity-law-of-the-peoples-republic-of-china-effective-june-1-2017/

Data Center Map—Colocation, Cloud and Connectivity. (n.d.). Retrieved July 2, 2024, from https://www.datacentermap.com/

Datacenters World Map. (n.d.). Datacenters World Map. Retrieved July 2, 2024, from https://map.datacente.rs

De Bernardo Caro, de Lota Protein, a Jorge Burgos: Los lobbistas tras la millonaria licitación del Registro Civil. (2021, August 13). El Mostrador. https://www.elmostrador.cl/destacado/2021/08/13/de-bernardo-caro-de-lota-protein-a-jorge-burgos-los-lobbistas-tras-la-millonaria-licitacion-del-registro-civil/

Diario Financiero. (2018). *Huawei detalla avances de Fibra Óptica Austral y cable alista viaje desde China a Chile*. Diario Financiero. https://www.df.cl/empresas/telecom-tecnologia/huawei-detalla-avances-de-fibra-optica-austral-y-cable-alista-viaje

Echevarría, R. (2020). *Infraestructura de Internet en América Latina: Puntos de intercambio de tráfico, redes de distribución de contenido, cables submarinos y centros de datos* (LC/TS. 2020 226; Desarrollo Productivo). Comisión Económica para América Latina y el Caribe (CEPAL).

Ellis, R. E. (2022). El Avance Digital de China en América Latina. *Revista Seguridad y Poder Terrestre*, *1*(1), 15–39. https://doi.org/10.56221/spt.v1i1.5

Eriksson, J., & Giacomello, G. (Eds.). (2007). *International Relations and Security in the Digital Age*. Routledge. https://doi.org/10.4324/9780203964736

Ex-Ante. (2021, November 16). *El documento del FBI contra la empresa china a la que se retiró la adjudicación de los pasaportes chilenos tras presión de EEUU.* Ex-Ante. https://www.ex-ante.cl/el-documento-del-fbi-contra-la-empresa-china-a-la-que-se-retiro-la-adjudicacion-de-los-pasaportes-chilenos-tras-presion-de-eeuu/

Flor-Unda, O., Simbaña, F., Larriva-Novo, X., Acuña, Á., Tipán, R., & Acosta-Vargas, P. (2023). A Comprehensive Analysis of the Worst Cybersecurity Vulnerabilities in Latin America. *Informatics*, *10*(3), 71. https://doi.org/10.3390/informatics10030071

G., D., & O., A. (2023, March 8). *Interview with Authorities at the Telecommunications Agency, Chile.* [In Person].

G. Z., A. (2023, March 13). *Interview with an Inter-American Development Bank Representative* [In Person].

Gayo, M. (2023, February 27). *LACNIC Blog | Status of Connectivity in Latin America.* LACNIC Blog. https://blog.lacnic.net/en/interconnection/status-of-connectivity-in-latin-america

General Data Protection Regulation (GDPR) – Legal Text (2016). (EU) 2016/679, OJ L 127, 23.5.2018. https://gdpr-info.eu/

Global Data Center Map. (n.d.). Retrieved July 2, 2024, from https://baxtel.com/map

Government of Argentina. (2018). *Argentina.gob.ar.* https://www.argentina.gob.ar/

Government of Argentina. (2019). *Argentina.gob.ar.* https://www.argentina.gob.ar/

Hirose, Y., & Toyama, N. (2020). *Chile Picks Japan's Trans-Pacific Cable Route in Snub to China.* Nikkei Asia. https://asia.nikkei.com/Business/Telecommunication/Chile-picks-Japan-s-trans-Pacific-cable-route-in-snub-to-China

Huawei Cloud Regions: Role, Positioning, and Division Strategy. (n.d.). Retrieved June 30, 2024, from https://forum.huawei.com/enterprise/en/huawei-cloud-regions-role-positioning-and-division-strategy/thread/691736221413425152-667213860102352896

International Finance Business Desk. (2020, September 21). *The Transoceanic Cable Project Is Big for Chile.* International Finance. https://internationalfinance.com/magazine/telecom-magazine/the-transoceanic-cable-project-is-big-for-chile/

Lode, S. L. (2019). "You Have the Data"...The Writ of Habeas Data and Other Data Protection Rights: Is the United States Falling Behind? *Indiana Law Jounal*, *94*(5), article 3. https://www.repository.law.indiana.edu/ilj/vol94/iss5/3

M. L., R. (2023, March 16). *Interview with a Former High-level Politician, Colombia.* [Online].

McGeachy, H. (2022). The Changing Strategic Significance of Submarine Cables: Old Technology, New Concerns. *Australian Journal of International Affairs*, *76*(2), 161–177. https://doi.org/10.1080/10357718.2022.2051427

Miller, J. B. (2020). *An Undersea Cable Deal with Chile Showcases Japan's Quiet Efforts to Counter China around the Pacific.* Business Insider. Retrieved June 24, 2024, from https://www.businessinsider.com/japan-undersea-cable-deal-chile-shows-engagement-in-latin-america-2020-10

More Subsea Cables Bypass China as Sino-U.S. Tensions Grow. (2024). Nikkei Asia. Retrieved May 15, 2024, from https://asia.nikkei.com/Spotlight/Datawatch/More-subsea-cables-bypass-China-as-Sino-U.S.-tensions-grow

O'Ryan, F. (2018, December 9). *China y Japón se disputan millonaria línea de fibra óptica a Chile y gobierno tomará decisión en 2019.* La Tercera. https://www.latercera.com/pulso/noticia/china-japon-se-disputan-millonaria-linea-fibra-optica-chile-gobierno-tomara-decision-2019/437537/

OAS. (n.d.). *Data Protection > Department of International Law > OAS.* Retrieved June 30, 2024, from https://www.oas.org/dil/data_protection_privacy_habeas_data.htm

Office of the United States Spokesperson. (2024). *Welcoming the First Subsea Cable between South America and the Indo-Pacific Region.* United States Department of State. https://

www.state.gov/welcoming-the-first-subsea-cable-between-south-america-and-the-indo-pacific-region/

P., M. (2023, February 13). *Interview with an Inter-American Development Bank Consultant* [In Person].

Park, S., García Zaballos, A., Iglesias Rodriguez, E., Puig Gabarró, P., Choi, K., Oh, D., Song, H., & Lim, J. (2022). *Roadmap for the Establishment of a Big Data Analysis Center for Critical Infrastructure Protection in Latin America and the Caribbean Post COVID-19*. Inter-American Development Bank. https://doi.org/10.18235/0003930

Putnam, R. D. (1988). Diplomacy and Domestic Politics: The Logic of Two-Level Games. *International Organization, 42*(3), 427–460. https://www.jstor.org/stable/2706785

Ragan, S. (2012, March 28). *PLA Concerns Lead to Huawei Being Blocked in Australia and Questioned in New Zealand.* SecurityWeek. https://www.securityweek.com/pla-concerns-lead-huawei-being-blocked-australia-and-questioned-new-zealand/

Ryngaert, C., & Taylor, M. (2020). The GDPR as *Global* Data Protection Regulation? *AJIL Unbound, 114*, 5–9. https://doi.org/10.1017/aju.2019.80

S., J. (2024, June 3). *Telecommunications Consultant with Direct Connections with the International Market* [Online].

Saravia, C. (2022). *Caso pasaportes: Registro Civil cobra a Aisino boleta de garantía y la acusa de incumplir las bases*. Diario Financiero. https://www.df.cl/empresas/industria/caso-pasaportes-registro-civil-cobra-a-aisino-boleta-de-garantia-y-la

Saravia, C., & Arteaga, M. (2021). *Embajada de China advierte por caso Aisino: "Puede suscitar preocupaciones sobre el entorno empresarial internacional en Chile"*. Diario Financiero. https://www.df.cl/empresas/industria/embajada-de-china-advierte-por-caso-aisino-puede-suscitar

Schulz, C.-A., & Rojas de Galarreta, F. (2020). Chile as a Transpacific Bridge: Brokerage and Social Capital in the Pacific Basin. *Geopolitics*. https://doi.org/10.1080/14650045.2020.1754196

Serentschy, G. (n.d.). *Digital Infrastructure Resilience and Security Policy Implications and Mitigation Measures*. Serentschy Advisory Services. Retrieved June 24, 2024, from https://www.serentschy.com/digital-infrastructure-resilience-and-securitypolicy-implications-and-mitigation-measures/

Submarine Cable Map. (n.d.). Retrieved June 24, 2024, from https://www.submarinecablemap.com/

Submarine Networks. (2024). *Humboldt Cable*. Submarine Networks. https://www.submarinenetworks.com/en/systems/trans-pacific/humboldt-cable

SUBTEL Chile. (2019). Public Call for Offers on the "Project Economic, Technical, and Legal Feasibility Study of the Submarine Cable Project: Asia-South America Digital Gateway." https://www.subtel.gob.cl/wp-content/uploads/2019/07/Specifications_SUBTEL_CAF_english.pdf

T. P. (2023, February 27). *Interview with a Former Economic Relations' Public Official, Chilean MOFA*. [Online].

Ting-Fang, C., Li, L., Suruga, T., & Tabeta, S. (2024). *China's Undersea Cable Drive Defies U.S. Sanctions*. Nikkei Asia. https://asia.nikkei.com/Spotlight/The-Big-Story/China-s-undersea-cable-drive-defies-U.S.-sanctions

WIC. (2022). *Introduction to World Internet Conference*. https://www.wicinternet.org/2022-09/22/c_814844.htm

Z., P. (2024, June 14). *Interview with a Security Consultant Focused on Latin America* [Online].

8 Surveillance and Facial Recognition

Unpacking Latin America's Complex Landscape

Surveillance technology poses ethical challenges for both private and public providers. The balance between privacy and public security is challenging to navigate.[1] The book *1984* and the concept of "Orwellian" are often mentioned in discussions about the consequences of government and corporate use of new technologies to monitor individual behavior in both digital and analog realms. This comparison is not accidental, and it stems from the general belief that surveillance, when misused, can give rulers and technology holders omnipresent power over individuals.[2]

Due to increased competition and backwardness advantage—discussed in earlier chapters—new surveillance technologies are also cheaper, easily replicable, and don't require human input to the same extent as it was needed before the introduction of artificial intelligence.[3] Surveillance methods, including collecting information and data, using security cameras (closed-circuit television [CCTV]), and monitoring digital behavior, such as retaining internet traffic, have significantly enhanced the government's ability to monitor people. This is despite efforts to minimize the negative impacts of these new forms of control, which have encroached on individuals' personal space.[4]

Biometric data includes all personal data "relating to the physical, physiological or behavioural characteristics of a natural person."[5] Biometrics can be used for identity verification, as is the case in border control, and for identification. Verification determines identity by comparing one person to themselves, while identification compares one person to a poll of images or a database.[6]

Facial recognition, a form of biometric data, is a crucial example of said surveillance innovations. Its use began to increase in the early 2010s[7] and is one of the most rapidly developing uses for artificial intelligence for security.[8] One of the main objectives of surveillance is the prevention of undesired social behavior or the violation of laws,[9] and facial recognition constitutes an ideal tool to that end, and also to monitor the transit of people across countries better and to categorize profiles.[10] While there are several systems in which facial recognition can be applied, they all share specific elements. They capture or integrate images, incorporate software to process them, and a database with reference "faces" the system can analyze. These systems—and the algorithms that discriminate and make decisions—are as good as the information they can compare, and the images used

DOI: 10.4324/9781003489450-10

need to fulfill a minimum quality standard. When those criteria are not met, the system cannot function correctly, leading to arbitrary discrimination and results. This is even more sensitive considering the data source is people's identities.[11] Considering all the features mentioned above, it is reasonable to assert that using facial recognition tools, while widespread, is contentious.

Apart from the blatant ethical considerations over privacy—racial, ethnic, and gender-based forms of bias are prevalent in both public and private systems, which specifically affect vulnerable social groups.[12] This is due to the nature of the technology, which requires a significant input of faces and struggles to identify features in those populations with less representation and quality data to train the algorithms and artificial intelligence behind the tools. The extension of this bias can cause issues such as the misidentification of suspects and the targeting of those members of society who are already considered vulnerable, as well as inefficiencies in the technology performance toward ensuring security in public areas.

In authoritarian countries, the use of these technologies can be directly linked to the state's capacity to maintain control over possible dissidents. The PRC's Social Credit System and CCTV facial recognition on the streets maintain a tight grip over social behavior by providing rewards and punishments that are measured through a score.[13] In the West, the increasingly important role of the private sector in the creation and diffusion of this form of technology has given companies real power related to the information garnered for public and commercial purposes. For example, governments have used social media searches to identify suspects.[14] In Europe, the use of CCTV and other surveillance technology varies greatly depending on the country, but most countries share common limitations and similar legal frameworks.[15]

Regarding security and international relations, the transfer of surveillance technology is highly controversial. Biometrical data is not only sensitive for individuals but can potentially be used as leverage or as an essential source of information to be analyzed in case of an international conflict, making it a crucial technology for states to maintain control over. The use of public data by other states—which I discuss in the next chapter—has had political and legal consequences in the US, Australia, and Europe. A clear example is the recent security discussions over TikTok and the potential of the nontransparent data transfer back to its parent company.[16] In terms of surveillance in Latin America, two examples could stand out. The first is the Passport and National ID public tender in Chile,[17] and the second is the meteorological balloon that crossed Colombia and increased awareness of the surveillance capacity of Chinese technology in South America.[18]

Considering this scenario, I suggest that facial recognition and surveillance technology are critical for analyzing how geopolitical tensions operate in Latin America and whether other actors actively participate in—and even dominate—the government contracts market.

At the domestic level, there is growing concern that these technologies could exacerbate power imbalances between governments and citizens[19] and that they considerably reduce privacy. This is expected in authoritarian governments, can be problematic for democracies, and can be a determinant for hybrid and weak

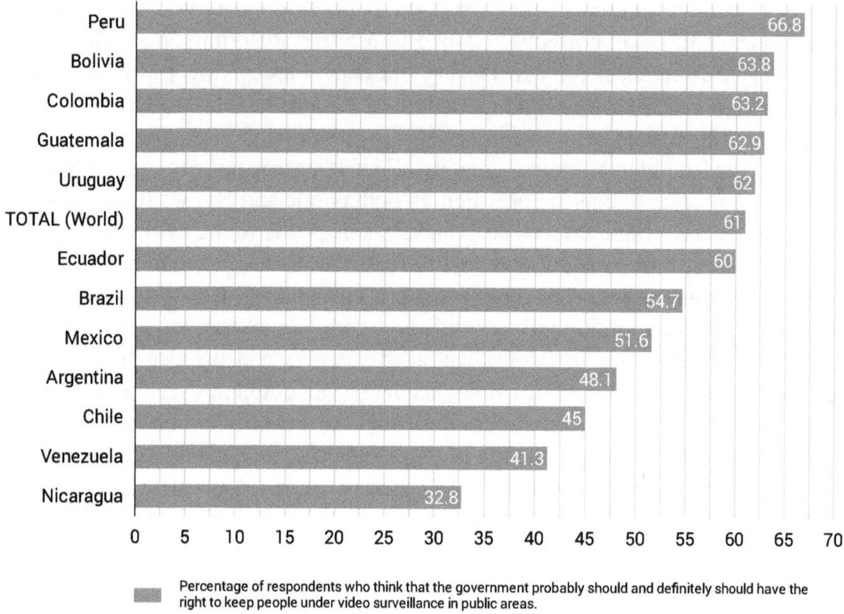

Figure 8.1 A graph comparing the percentage of respondents of each target country who think the government should have the right to keep people under video surveillance in public areas. (World Values Survey, Haerpfer et al., 2022, Q.196.)

democratic states to shift to authoritarian practices to ensure control under the justification of technologies. In Latin America, where democratic levels have decreased and citizens are dissatisfied with their governments, this is an ongoing concern.[20]

Latin American citizens' attitudes toward government surveillance exhibit significant variation. The World Values Survey (WVS) considers individuals' relationship with technology and security through several questions. When asked if the "Government has the right: Keep people under video surveillance in public areas," as shown in Figure 8.1, 61% of the world respondents answered that it probably should or should have the right to do so. Countries like Colombia, Uruguay, Peru, and Bolivia exhibit a similar trend. In comparison, only 54% of Brazilians, 52% of Mexicans, 48% of Argentinians, 45% of Chileans, and 41% of Venezuelans support video surveillance.[21]

A second question regarding the "Government has the right: Collect information about anyone living in [COUNTRY] without their knowledge" also reflects this difference. About 28% of international respondents believed that their government probably or definitely should have the right to collect data. As Figure 8.2 shows, while more than 30% of citizens in Argentina, Bolivia, Ecuador, or Guatemala support this practice, Brazil, Chile, Uruguay, and Venezuela respondents were much more skeptical of unrestricted digital surveillance.[22] Interpretations of these

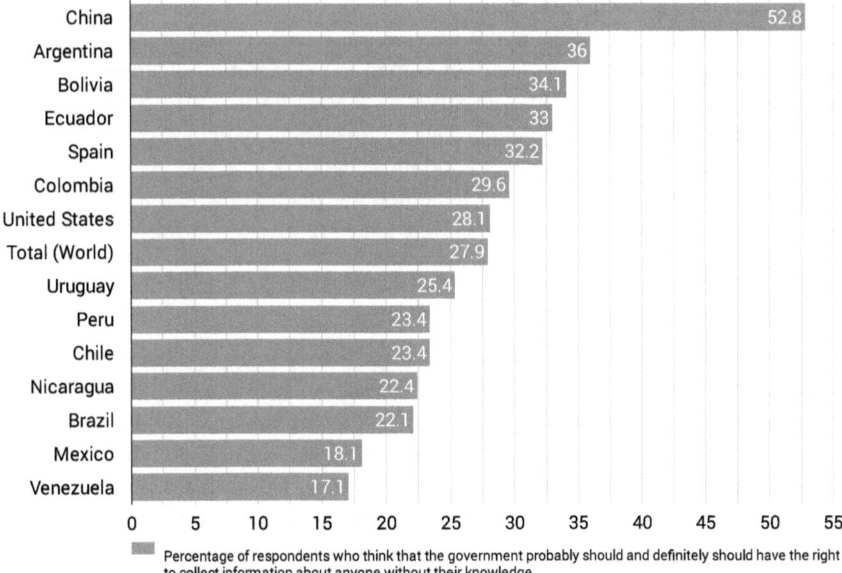

Figure 8.2 A graph comparing the percentage of respondents of each target country who think the government should have the right to collect personal data. (World Values Survey, Haerpfer et al., 2022, Q.198.)

attitudes can likely be linked to historical phenomena, such as authoritarianism in recent history or violent insurgencies and cartels.[23] Regardless of the reasons, it is evident that all states in Latin America employ facial recognition in public services or for public security to some extent. Consequently, the debate over the limits and uses of this technology and the nature of the providers of surveillance tools has intensified in recent years.

The variation in attitudes toward surveillance and data collection may influence how Latin American countries engage with global technology providers and shape their diplomatic relations with key international players. In some cases, controversial actors such as Chinese, Russian, and Iranian providers have supplied facial recognition systems in Latin America.[24] China's expanding role in providing surveillance technology to Latin American countries has raised concerns about digital sovereignty and the potential for increased state control over citizens, and also the direct access the CCP would have to the data used in those surveillance systems due to the Chinese Cybersecurity Law and the government's tight grip of the private sector and their businesses overseas.[25] Conversely, the US—which companies don't often participate in surveillance public tender competitions—has advocated for stricter data protection measures and transparency in surveillance technologies, emphasizing the importance of safeguarding civil liberties.

In this chapter, I explore the state of surveillance technology, particularly facial recognition, in Latin America. I provide an overview of some of the leading companies involved in this market, the behavior of states, and possible controversies related to sensitive forms of digital security measures. I present the situation in Brazil and Argentina as two key and often-mentioned examples of advanced implementation of facial recognition initiatives at a state level. I also explore several cases in which Chinese public and private companies have helped provide surveillance technology to governments, including Ecuador, Cuba, and Venezuela. I analyze these cases to understand this phenomenon's geopolitical implications and to portray better how values and norms might shape technology diffusion outputs.

Latin America's Use of Facial Recognition

Latin American states predominantly depend on private enterprises to deliver facial recognition services nationally.[26] However, national and foreign state-owned companies compete to provide governments with surveillance systems. The mechanisms for awarding these contracts differ across countries, ranging from public tender processes to direct negotiations and donations within Official Development Assistance (ODA) or grants. The applications of these technologies are also extensive, with a range of uses from passports to CCTV) cameras that integrate facial recognition capabilities.[27]

Relying on private entities for critical surveillance infrastructure raises essential questions about governance, accountability, and privacy implications, which align with the argument and theory explored in this book. The varying methods of contract allocation reflect differing levels of transparency and regulatory oversight, which can impact public trust and the efficacy of these technologies. Furthermore, integrating facial recognition in diverse applications, from identity verification in passports to monitoring through CCTV, underscores the pervasive nature of this technology in public and private spheres.

Conversely, a report by Venturini and Garay provides a clear overview of the state-of-the-art technology in the region. They highlight that several companies currently providing facial recognition tools to states and institutions in Latin America provide evidence of participation of firms that have directly been questioned for contributing to human rights abuses. For example, the Chinese companies Dahua and Hikvision have been banned in the US but have secured contracts in Mexico. In addition, IDEMIA (formerly known as Morpho Safran) has also raised concerns among international organizations.[28] The French company, mentioned several times in this chapter, provides biometric technology to countries for applications ranging from passport identification to surveillance in both the US and Latin America. In the following chapter, I elaborate on data-collection and data-management technology, exploring cases in which Beijing and Washington have attempted to push for certain forms of development. I also discuss a specific case in which IDEMIA was selected over a Chinese provider to manage Chile's ID databases and provide the country's passport system. Venturini and Garay complement their report with a platform dedicated to mapping the presence of contracts related to facial recognition in the region. According to the site, companies' origins include European countries, the US, Korea, and China. Still, there is a relevant presence of companies of Latin American origin, predominantly from Argentina, Brazil, Colombia, and Costa Rica.[29]

Countries in Latin America utilize these technologies for purposes similar to those of the rest of the world. Governments value introducing tools that can help maintain order and security and identify possible suspects or antisocial behavior on the streets. The public's receptiveness to facial and biometric technology varies depending on the state of political affairs, the record of violence, and the history of each nation. For example, considering the incidence of guerrilla violence, I argue that countries like Colombia are expected to welcome enhanced security. The same can be said about Peru due to its violent history of terrorist attacks by the Sendero Luminoso. On the other hand, Chile might be more hesitant due to the prejudice against security agencies that have derived from the traumatizing history of authoritarianism from the 1970s through the 1990s.[30]

However, a few cases stand out regarding the implications of implementing these technologies for this book. That is the regional production of technology and relevant cases of import of surveillance tools from outside Latin America. In that sense, several actors highlight that Argentina and Brazil are two of the most developed countries in the region regarding facial recognition tools and their use in public spaces.

Argentina's Cuban Biometric System and Broad Facial Recognition Market

Argentina, a federal nation with approximately 46 million inhabitants, faces the challenge of adequately tracking and identifying its citizens. In addition to the limitations given by its extensive geography and somewhat fragmented political system, the country has experienced episodes of violence, including two major terrorist attacks in Buenos Aires during the 1990s. The first attack took place at the Israeli Embassy in 1994, and the second targeted the Jewish cultural center AMIA, resulting in the loss of 85 lives and over 300 injuries. These cases, later potentially connected to Hezbollah and the Iranian government—in a series of investigations that also involved the death of prominent persecutor Alberto Nisman—potentially shaped domestic political opinion toward the strengthening of security and control within the borders.[31] Since the early 2000s, legislation has been passed to implement widespread surveillance and security measures in the country.[32] However, the solutions that the government has resorted to were even described by Julian Assange as "the most aggressive surveillance system in all of Latin America."[33]

In 2009, the Cristina Fernandez de Kirschner (CFK) administration issued a decree to modernize the national ID system, which culminated in 2011 with the 2011 "Decree 1,766" and the creation of the *Sistema Federal de Identificación Biométrica para la Seguridad* (SIBIOS) or the federal system of biometric identification for security purposes. The system would gather, sort, and process up to 37 million citizens' biometric data, including fingerprints and facial feature recognition, by 2012. In the following years, updates to the system included incorporating a "smart chip" to interact with the national citizens registry (RENAPER).[34]

SIBIOS is controversial due to the power it gives the incumbent government over citizens' biometric data, as well as the nature of the providers of this technology. The two leading providers by 2021 were IDEMIA (formerly Morpho Safran), a French company later discussed in this chapter, and DATYS, a Cuban firm.[35] It is not a secret that Cuban businesses have been heavily controlled by the state,[36] and until 2020, all mixed companies required over 50% of shares to be owned by the Cuban government.[37] The Castrista regime, which has been in government for more than 60 years, is internationally known to be strictly authoritarian and to provide support to other left-wing—both democratic and authoritarian—governments in Latin America.[38]

DATYS was founded in La Habana in 2005, has around 700 employees, and is focused on biometrics, identity, security, management, and data mining. According to a report by the Asociación por los Derechos Civiles (Association for Civil Rights), the technology imported from Cuba allows the Argentinian government to work with facial images to reinforce an automatic identification mechanism. Nevertheless, the central identification system used by RENAPER is through fingerprints.[39]

Considering this, I posit a direct correlation exists between the *Castrista* regime and the implementation of SIBIOS in Argentina. This is supported by a series of open statements by Argentinian leaders over the years, particularly by the Kirschner and Peronist factions, that emphasize the role Cuba had in the development of the system since the island was among the first countries in the region to implement biometric identification.[40] In 2011, president CFK publicly thanked the Republic of Cuba for its collaboration in developing low-cost software to integrate into the AFIS system to identify people in the country. In 2015, the Argentinian Interior Ministry communicated that it would continue cooperating with Cuba to develop facial recognition systems.[41] Finally, Morpho—now IDEMIA—is said to have provided the Argentinian government with tools that communicated with the identification system, enabling effective control of biometric data.[42] The company has denied that it offered governments in Brazil, Ecuador, and Argentina surveillance technology after being openly questioned about human rights concerns related to this technology. In response to those accusations, they stipulated that they provide control for stadiums, fingerprint technology, and support for judicial investigations. Moreover, it declared it did not provide the Argentinian government with facial recognition tools.[43]

Although surveillance can be justified to maintain order and ensure public safety, the initial international and domestic reaction to the Argentinian initiative is the legal apparatus surrounding it. This apparatus enables the exploitation of databases that allow facial recognition through RENAPER and surveillance cameras, as well as shared biometric data, shrinking individual privacy.[44] In some cases, the police have used video recognition to look for fugitives by comparing live footage against a national register of facial pictures.[45]

Argentina also utilizes facial recognition at regional and local levels of government administration. For example, in Buenos Aires, a suspect recognition system, "Sistema de Reconocimiento Facial de Prófugos (SRFP)," has been in operation since 2019, provided by the Argentinian company DANAIDE. The Japanese firm NEC was hired to maintain and provide hardware for a facial recognition system in the Tigre area.[46] The Korean company Nubicom was selected to implement a facial recognition system in the city of Salta.[47] Israel-based companies have also ventured into the facial recognition market in Latin America. Notably, Cellebrite and AnyVision have been accused of facilitating government surveillance and have provided regional countries such as Argentina with technologies in this area.[48]

Argentina is, therefore, an attractive market for surveillance technology companies. Something interesting about this case is the apparent absence of American and Chinese companies competing to enter the facial recognition market. One outcome of this trend is that the international reputation of Korean, Japanese, and European companies makes them a more "neutral" alternative to providing biometric information software and hardware. However, the case of Cuban technology could be regarded as "more controversial" if technologies embed certain ethical and political traits related to their countries of origin.

Brazil: A Leader in Private Facial Recognition Implementation?

Brazilian companies are directly involved in developing and supplying biometric technology, making them direct competitors of other relevant actors in the field. Although different countries in the region also produce this tracking technology, the scale of the Brazilian surveillance market, the country's demographics—more than 215 million people—and history make it a relevant case to compare. More than 30 Brazilian cities have implemented facial recognition to some extent, and its use goes from preventing fraud to tracking school attendance and even marketing purposes.[49] In 2018, ViaQuatro—a Brazilian company—was in the spotlight because of the processing of images with facial recognition software without consent in the São Paulo subway system for advertising purposes instead of maintaining safety in the public system. *The Instituto de Defesa de Consumidores (IDEC), or the Consumer Defense Institute,* issued a complaint culminating in a $500,000 Brazilian Real fine for the company.[50]

Relevantly, SERPRO, a public company linked with the Brazilian Finance Ministry, oversees the processing and managing of all forms of biometric data. Alleged misuse of personal data has resulted in two judicial initiatives against the institution. One is because of undisclosed data transfers between the company and the Brazilian intelligence agency (ABIN), and the other is because of using a driving license database to offer facial recognition services.[51]

Foreign companies also provide Brazil with facial recognition technology, sometimes directly or sometimes in collaboration with local companies that purchase and adapt their technology. Admobilize, an American firm, was selected to provide facial recognition to São Paulo's underground system for several months in 2018. FaceWatch, a British company, provided a private Brazilian firm with a facial recognition system for the São João festival in the Paraíba province in 2019. This system was still operating in 2022 when they were used to arrest 25 people during that year's festival. In the São Paulo metro system, Engie Ineo Johnson, a consortium of French, Brazilian, and American companies, was selected to provide the facial recognition system.[52] Another example is IDEMIA (formerly Morpho Safran), as the French company tested a new facial recognition system at several Brazilian airports in 2021.[53]

Despite its recognized usefulness in criminal cases—even more so in countries with a considerable population and crime rate—facial recognition in Brazil has also been criticized for its potential bias and racial discrimination. The police department of the state of Ceará, which uses this type of system to identify suspects, was in the spotlight when its software failed to identify black faces and put Michael B. Jordan, an African American actor—famous for his role in Black Panther, among the most-wanted criminals after a tragic shooting on Christmas Eve in 2021. Since racism is an ongoing and sensitive issue in Brazil, these cases enhance public skepticism toward the private and public use of technology.[54]

Considering the analysis of the most relevant controversies surrounding facial recognition use in Brazil, one common criticism is the lack of transparency regarding the processing and use of the data. There is also an ongoing discussion on overseeing institutions and regulators' role in properly developing facial recognition as a state tool to ensure security.[55] A 2024 study by the Brazilian Data Protection Authority (ANPD) highlighted the need for security and control that facial recognition technology brings. Still, there are risks for individuals regarding potential discrimination and improper data use.[56] However, the origin of the supplying companies or the broader geopolitical considerations in using facial recognition technology are only sometimes mentioned in most policy debates in the country. In this sense, the case of Brazil can be indirectly related to the overarching puzzle of this book, given the focus on domestic companies involved in the production of the technology and the participation of American and European companies in the Brazilian market.

Chinese Increased Interest in Providing Facial Recognition Tools

Beijing has continually sought to promote surveillance technology as a pillar of the perceived security of the PRC. As some Latin American countries have high crime rates, it is unsurprising that access to cheap and direct control tools can be attractive to the general population, which is reflected in the support for surveillance in some countries as measured by the World Values Survey. The strategy of promoting a narrative that associates "security" with "China" is readily apparent, and examples of these efforts range from local Chinese business associations to large-scale government negotiations. In 2024, a 12-meter Chinese-style gate was installed at the entrance to Chile's *Meiggs* neighborhood in Santiago, traditionally associated with Chinese migrants. The goal was to increase the perception of security, which was sponsored by the local Chinese Trade and Culture Association and supported by the Chinese embassy and the Santiago municipality.[57] The notion of a "Safe China" is also shared by some members of the Latin American elite and is widely shared among businesspeople who trade or negotiate with Chinese companies and institutions.

It is not a surprise that Chinese technological and surveillance companies have made a debut in Latin America. However, their market entry method differs from long-established European and regional firms. In the facial recognition database of "identified vendors," there are three companies that have allegedly participated in public surveillance systems using facial recognition: Hikvision, Dahua, and Huawei—all of which have faced restrictions in the US either for security reasons or their involvement in human rights abuses in China.[58]

Dahua provided the Mexican state of Coahuila with a modern video surveillance system meant to cover the whole region with cameras for crime prevention and identification.[59] The decision was sensitive since Dahua had been blacklisted for its connections to privacy violations and persecution in China. It was also called out by the Business & Human Rights Resource Centre to respond regarding a report accusing the firm of deploying surveillance technology in Latin America "without transparency or public scrutiny eroding democratic processes, undermining privacy, freedom of speech and other basic human rights," to which it did not respond.[60] In Brazil, Dahua also provides security cameras and other forms of surveillance technology to the states of Caerá and Espíritu Santo. In both cases—and despite the Bolsonaro Government being openly critical of the PRC—the contract was granted through direct competition.[61]

A similar case happened in the city of Pico Truncado in México, which selected Hikvision as its security provider.[62] In Panama, the company Huawei was chosen to provide the technology for a Center of Operations for Security and Emergencies (C2)—the product of a direct contract between China and the Panameñan state, only a year after the latter changed its diplomatic recognition from Taiwan to the PRC.[63] This could be interpreted as a direct result of the benefits the country was looking forward to receiving as part of this historic decision.

Other cases of implementation of Chinese surveillance technology also involve direct negotiation. Ecuador has received 6,000 security cameras from China as part of their co-developed security system called "ECU-911." This initiative began in 2008 when an Ecuadorian delegation visited China during the Olympic games that summer; in 2011, a commission visited Beijing to define the terms of the agreement.[64] The negotiations derived into a 240 million dollar loan provided by China Development Bank (CDB),[65] and implementation was divided into several steps, culminating not only in the installation of surveillance technology but also in 16 on-site centers to process emergency calls and operate cameras—employing over 3,000 people.[66] As for the origin, the hardware and software used in its operation mostly come from Chinese companies, but there are also examples of providers from other origins. Some data storage devices were acquired from Hewlett Packard (HP),[67] and security cameras from VERINT,[68] both American companies. Interestingly, Ecuadorian providers almost did not take part in this endeavor.[69]

The installation of this system is said to have reduced the crime rate and was initially well-received—with overall analysis being positive on the consequences for Chinese–Ecuadorian relations.[70] Nevertheless, there are criticisms of using the data garnered and processed through video, facial recognition, and surveillance tools in general. First, the government in power during the negotiation and implementation of this system—the Rafael Correa administration—had a track record of populism and pressure over the opposition.[71] Second, a report by the *New York Times* highlighted the potential risks to the population and human rights violations that the security apparatus could commit. The main problem for local activists is the lack of due diligence and the uneven playing field between the two states, which makes it almost impossible for Ecuadorians to reach their Chinese counterparts to request information.[72] Following the same line, Bolivia—under Evo Morales' administration—also negotiated with Beijing to install a security system named BOL-110. The system was repurposed during COVID-19 quarantines to follow potential infected citizens.[73] Again, criticism from civil society stemmed from the fact that the facial recognition system and digital surveillance granted the government unsupervised access to citizens' privacy, including their movement around the cities where the system was implemented.[74]

Chile has also received support from China to enhance the national security system. However, there hasn't been much discussion on facial recognition—as is the case in other countries—but rather on the use of data and geopolitical issues.[75] However, in 2024, several members of Congress were sponsored by the Chinese state to visit Shenzhen—the so-called Chinese Silicon Valley—to learn about artificial intelligence but mainly focused on identifying missing persons.[76]

China's share of the facial recognition market in Latin America has been growing slowly but steadily. In most cases, contracts are obtained through direct negotiations, taking advantage of reputational factors and China's ability to provide loans and assistance to developing countries. In other words, these companies are entering the market with the direct support of Chinese institutions. Regarding security implications, the debate over the ownership and potential use of surveillance data has been rather lackluster, and civil society has focused more on potential state violations of privacy and persecution than on a broader geopolitical debate. Although some Chinese companies have been blacklisted in the US,[77] this has not prevented them from reaching local and national agreements with Latin American governments.

Conversely, the debate on these companies' activities in China's surveillance apparatus—and human rights violations—do not get much public attention, nor the implications of a rivalry between Washington and Beijing. I interpret this as a clear advantage for China in its attempt to transfer technology to the region and gain a better reputation as the origin of technology providers. However, by exporting security systems that are used in the systemic persecution of minorities and repression, there is an argument on whether these tools can effectively allow Latin American states to become more authoritarian or begin to share some of China's security rationale and values.

American Facial Recognition and Resistance to Chinese Technology

Facial recognition technology has been framed as a critical security measure in the US. The 9/11 terrorist attacks had a profound influence on policy and public sentiment, paving the way for the implementation of security protocols with minimal resistance. Proponents suggest employing more sophisticated surveillance technology could have averted these tragic events.[78] However, debates have arisen due to the use of facial recognition technology, particularly regarding its unacknowledged or unprepared implementation by security agencies,[79] racial profiling,[80] or the use of this technology in undocumented migrants. Hence, it is not counterintuitive to conclude that the attitude toward facial recognition in the US is divided between those concerned and those who accept it.[81] Some well-recognized American companies have begun to provide facial recognition systems to the public sector. For example, Amazon has provided law enforcement in the US with this technology[82] and started tests in the United Kingdom's train stations.[83]

In Latin America, I found that the presence of companies from the US has been limited, and their participation is not as controversial. VERINT is a US-based company with significant operations in Israel that provided several Ecuadorian institutions, including ECU-119 and the city of Guayaquil, through a distributor called Union Electrica S.A. This company has been questioned for its role in providing the governments of Kazakhstan and Uzbekistan with surveillance, control, and repression capacity.[84] Another exception could be the Brazilian case, including debates about the use of AdMobilize's advertising face recognition in the subway system, and the "Consórcio Engie Ineo Johnson," which brought together French, Brazilian, and American companies to manage subway security using face recognition.[85] I foresee that American technology might be used in the private sector or public places operating with private security. Still, the decision-making process and the objectives behind these technologies' application differ from the cases reviewed in this chapter.

Regarding political action, Washington has taken countermeasures against the presence of Chinese companies in the domestic facial recognition market.[86] Examples include the blacklisting of firms such as Megvii Technology, Hikvision, and Dahua because of their involvement in human rights abuses in Xinjiang, contributing to the repression and genocide against the Uyghur people.[87] Through its embassies, the US has disseminated the reasoning and recommendations regarding pushback against Chinese surveillance technology companies—and the development of military technology,[88] which shows an intent to raise awareness in Latin American countries regarding using PRC tools for surveillance.

Conclusion: Analyzing the Impact and Implications of Facial Recognition Technology in Latin America

In this chapter, I explored the state of surveillance technology, particularly facial recognition, in Latin America. I detailed the involvement of various international and domestic companies, highlighting how their technologies are integrated into public and private sectors. This exploration considered the ethical concerns and potential biases that arise with facial recognition, such as privacy issues, racial and gender bias, and the broader societal implications of such technologies. This is crucial to the analysis since the deployment of facial recognition technology often involves extensive data collection, raising concerns about privacy and potential abuse. Moreover, racial recognition systems can perpetuate existing biases, mainly if the data used to train these systems are not always representative, leading to discrimination in the system's response. I also contextualized that the public's perception of facial recognition technology varies significantly across the region. This variation is influenced by historical experiences, current political climates, and the transparency of the processes surrounding the technology's implementation.

Considering the delicate nature of biometrics in the public and private sectors, I addressed the potential for misuse of facial recognition technology, such as in authoritarian regimes where it could suppress dissent. The ethical concerns were tied to broader questions about governance, accountability, and the potential for power imbalances between state authorities and citizens. Although these are global concerns, the divided opinions on surveillance, the partisan use of these technologies, and the lack of transparency—in some cases—have brought up debate in Latin America.[89]

After conducting the literature review to determine relevant actors and case studies, Brazil and Argentina appeared to be prime examples of advanced implementation. These examples illustrated the varying levels of technological adoption and the distinct approaches taken by different countries in Latin America. I also described the participation of companies from the US, China, France, and Korea in the Latin American market. I discussed how these companies' products are employed in various sectors, from public safety to private security, and the different levels of state engagement, ranging from active participation to regulatory oversight.

Following this book's objectives, the analysis sought to understand the geopolitical implications and how values and norms might shape technology diffusion outputs. My first argument regarding this issue is that having foreign companies manage sensitive biometric data can be a security challenge for countries when there is not a clear framework to safeguard the use of that information. For this reason, private actors that use and process facial recognition technology must provide states with reputational and technical capital or be highly monitored and questioned. Some countries, mainly middle powers such as Korea, Japan, or France, seem to be less controversial in the supply of facial recognition technology. Consequently, some of the cases involving Chinese companies have become sensitive because of the involvement of these providers with accusations of human rights violations. However, I found that outside of that normative and political context, debate on geopolitical affairs or the more considerable competition between the US and the PRC is developing and somewhat limited. Instead, regional civil society organizations and individuals critical of facial recognition have focused on the impact of these technologies on privacy and the lack of adequate data protection. There are even movements in Latin America to ban these tools because facial recognition is incompatible with fundamental rights, including privacy. Legislation does not adequately protect citizens in this regard, and it will exacerbate existing prejudices.[90]

In other words, despite the sensitive nature of facial recognition technology and its association with security analyses, the Latin American debate on its application has been somewhat connected to governments' limitations on citizens' privacy and the implications of widespread facial recognition for democracies. A handful of these interpretations state that Chinese companies are enabling the regional decline of democratic institutions and greater state control by transferring this technology, which can be interpreted as a regional security concern.

As for specific measures, this analysis concludes that there are crucial pending tasks in Latin America regarding using and implementing facial recognition technology. First, developing robust regulatory frameworks to govern facial recognition technology to address data protection, transparency, and accountability issues. Second, increasing public awareness and engagement around surveillance technologies can help build trust and ensure these systems are used responsibly. Third, given the geopolitical implications, international cooperation and dialogue are necessary to establish norms and standards for using facial recognition technology. This includes addressing concerns about transferring sensitive data and the potential for foreign influence. Finally, it is necessary to encourage technological innovation while ensuring ethical considerations by addressing biases in data and algorithms and ensuring that technologies are developed and deployed in ways that respect human rights.

Notes

1 Macnish (2017); Königs (2022).
2 Königs (2022); Lyon (2001), p. 174.
3 Königs (2022).
4 Lyon (2001), pp. 172–174.
5 European Union (n.d.).
6 Zureik and Hindle (2004), p. 115.
7 Caeiro (2022), p. 7.
8 Smith and Miller (2022), p. 167.
9 Lyon (2001), p. 174.
10 Smith and Miller (2022), p. 167.
11 Venturini and Garay (2021), p. 4.
12 Souza and Zanatta (2021), p. 2; Caeiro (2022); Smith and Miller (2022).
13 Creemers (2018), pp. 2–4.
14 Smith and Miller (2022), p. 174.
15 Dratwa (2014), pp. 20–23.
16 Hannig Núñez and Ichihara (2024).
17 Discussed in detail in the next chapter.
18 R. M. L., personal communication (March 16, 2023); Bubola (2023).
19 Smith and Miller (2022), p. 174.
20 Lagos (2023).
21 Haerpfer et al. (2022), Q198.
22 Haerpfer et al. (2022), Q198.
23 Bertoni and Kurre (2017), p. 326; Caeiro (2022).
24 Caeiro (2022), p. 16.
25 Creemers et al. (2017).
26 03 Facial Recognition Rollouts: Trends in Buenos Aires and São Paulo. (n.d.).
27 Venturini and Garay (2021).
28 Venturini and Garay (2021), pp. 16–17.
29 *Proveedores Identificados* (n.d.).
30 Bertoni and Kurre (2017), p. 326.
31 Caro (2013).

32 Bertoni and Kurre (2017).
33 Infobae América (2013).
34 Avaro (2014), p. 100.
35 Venturini and Garay (2021), p. 11.
36 Andarcia (2021).
37 France 24 (2020).
38 Querido (2021).
39 Asociación por los Derechos Civiles (2017), pp. 27–29.
40 Asociación por los Derechos Civiles (2017), p. 12.
41 Asociación por los Derechos Civiles (2017), p. 13.
42 Frescura Toloza (2019), p. 12.
43 *Response from IDEMIA to allegations about sale of surveillance technology in Latin America* (2022).
44 Rodriguez (2012).
45 Caeiro (2022), p. 11.
46 *Tigre lanzó NeoCenter, un sistema de vanguardia en reconocimiento facial* (2019).
47 Venturini and Garay (2021), p. 11.
48 Pisanu and Arroyo (2021), pp. 7–14.
49 Caeiro (2022), p. 12; Belli et al. (2024).
50 Faraco Cebrian et al. (2024), p. 10.
51 Venturini and Garay (2021), pp. 12–13; Belli et al. (2024).
52 Venturini and Garay (2021), pp. 12–13.
53 IDEMIA (2021).
54 Gonzalez Omerod (2022).
55 Faraco Cebrian et al. (2024), pp. 8–9.
56 Faraco Cebrian et al. (2024), pp. 8–9.
57 Carneiro (2024).
58 Aristegui Noticias (2020).
59 Government of Coahuila, Mexico (2020).
60 *Dahua Technology Did Not Respond to Allegations about Sale of Surveillance Technology in Latin America & Questions from CSO on Human Rights* (2022).
61 Majerowicz and Carvalho (2024).
62 *Nuevo sistema de Videovigilancia Urbana y Monitoreo en la ciudad de Pico Truncado* (2020).
63 Government of Panama (2018).
64 Morena-Álvarez et al. (2023).
65 *Project | China.Aiddata.Org* (n.d.).
66 Mozur et al. (2019).
67 Morena-Álvarez et al. (2023).
68 Pisanu and Arroyo (2021), p. 30.
69 Morena-Álvarez et al. (2023).
70 Alter et al. (2024).
71 Ellis (2022), p. 28.
72 Mozur et al. (2019).
73 Ellis (2022), p. 29.
74 Pelcastre (2019).
75 Discussed in the next chapter.
76 Wilson (2024).
77 U. S. Mission Chile (2021).
78 Gates (2011), pp. 2–3.
79 Johnson (2023).
80 Amnesty International (2022).
81 Kostka et al. (2021).

82 Dwoskin (2018).
83 *Amazon Facial Recognition Tests at UK Train Stations Enhance Security* (2024).
84 Pisanu and Arroyo (2021), p. 30.
85 *Metrô define empresa que vai fornecer sistema de reconhecimento facial por R$ 58,6 milhões* (2019).
86 Majerowicz and Carvalho (2024).
87 *US Blacklists China Organisations over Xinjiang "Uighur Abuse"* (2019).
88 U. S. Mission Chile (2021).
89 Bertoni and Kurre (2017), p. 339.
90 Caeiro (2022).

Bibliography

03 Facial Recognition Rollouts: Trends in Buenos Aires and São Paulo. (n.d.). Chatham House–International Affairs Think Tank. Retrieved July 16, 2024, from https://www.chathamhouse.org/2022/11/regulating-facial-recognition-latin-america/03-facial-recognition-rollouts-trends-buenos

Alter, J. S., Cook, J. A., & Dussel Peters, E. (2024). Chapter 13: What the ECU-911 Project Has Brought to Ecuador and to China-Ecuador Cooperation. In *Connecting China, Latin America, and the Caribbean: Infrastructure and Everyday Life* (pp. 334–358). University of Pittsburgh Press. https://muse.jhu.edu/pub/49/edited_volume/chapter/3896982

Amazon Facial Recognition Tests at UK Train Stations Enhance Security. (2024). https://aibusiness.com/computer-vision/amazon-facial-recognition-tests-at-uk-train-stations-enhance-security

Amnesty International. (2022). *Estados Unidos: La tecnología de reconocimiento facial refuerza la práctica policial racista en Nueva York.* https://www.es.amnesty.org/en-que-estamos/noticias/noticia/articulo/estados-unidos-lla-tecnologia-de-reconocimiento-facial-refuerza-la-practica-policial-racista-en-nueva-york/

Andarcia, M. V. (2021). *Cuba Legalises First Private Companies without Relaxing Political Oppression.* Global Affairs and Strategic Studies. https://en.unav.edu/web/global-affairs/cuba-legaliza-las-primeras-empresas-privadas-sin-relajar-la-opresion-politica

Aristegui Noticias. (2020). *Vigilancia biométrica: El tortuoso camino de Coahuila hacia el reconocimiento facial.* Aristegui Noticias. https://aristeguinoticias.com/1111/mexico/la-llegada-de-dahua-technology-a-mexico-pone-en-alerta-a-defensores-de-derechos-humanos-el-gobierno-de-coahuila-no-ha-transparentado-los-montos-pagados-a-la-empresa-china-foto-quinto-elemento-lab/

Asociación por los Derechos Civiles. (2017). *La identidad que no podemos cambiar: Cómo la biometría afecta nuestros Derechos Humanos* (pp. 1–32). Asociación por los Derechos Civiles. https://www.pensamientopenal.com.ar/system/files/2017/05/miscelaneas45320.pdf

Avaro, D. (2014). Citizen Traceability: Surveillance à la Argentina. *Journal of Power, Politics & Governance*, *2*(3 & 4). https://jppg.thebrpi.org/vol-2-no-3-4-december-2014-abstract-6-jppg

Belli, L., Gaspar, W. B., & Zingales, N. (2024). Regulating Facial Recognition in Brazil: Legal and Policy Perspectives. In M. Zalnieriute, & R. Matulionyte (Eds.), *The Cambridge Handbook of Facial Recognition in the Modern State* (pp. 228–241). Cambridge University Press. https://doi.org/10.1017/9781009321211.019

Bertoni, E., & Kurre, C. (2017). Surveillance and Privacy Protection in Latin America: Examples, Principles, and Suggestions. In F. H. Cate, & J. X. Dempsey (Eds.), *Bulk Collection* (1st ed., pp. 325–342). Oxford University Press. https://doi.org/10.1093/oso/9780190685515.003.0016

Bubola, E. (2023, February 6). Another Chinese Balloon Flew over Latin America, China Confirms. The New York Times. https://www.nytimes.com/2023/02/06/world/americas/chinese-balloon-latin-america-colombia.html

Caeiro, C. (2022). *Regulating Facial Recognition in Latin America: Policy Lessons from Police Surveillance in Buenos Aires and São Paulo*. Royal Institute of International Affairs. https://doi.org/10.55317/9781784135409

Carneiro, R. (2024, July 1). *Chinatown santiaguino: Barrio Meiggs inaugura pórtico chino con la finalidad de aumentar su seguridad*. BioBioChile - La Red de Prensa Más Grande de Chile. https://www.biobiochile.cl/noticias/nacional/region-metropolitana/2024/07/01/chinatown-santiaguino-barrio-meiggs-inaugura-portico-chino-con-la-finalidad-de-aumentar-su-seguridad.shtml

Caro, I. (2013). Los atentados de 1992 y 1994 en Buenos Aires: Sus repercusiones en las relaciones de Irán con Argentina. *Atenea (Concepción)*, *507*, 165–179. https://doi.org/10.4067/S0718-04622013000100011

Creemers, R. (2018). China's Social Credit System: An Evolving Practice of Control. *SSRN Electronic Journal*. https://doi.org/10.2139/ssrn.3175792

Creemers, R., Webster, G., & Triolo, P. (2017). *Translation: Cybersecurity Law of the People's Republic of China* (Effective June 1, 2017). DigiChina. https://digichina.stanford.edu/work/translation-cybersecurity-law-of-the-peoples-republic-of-china-effective-june-1-2017/

Dahua Technology Did Not Respond to Allegations about Sale of Surveillance Technology in Latin America & Questions from CSO on Human Rights. (2022). Business & Human Rights Resource Centre. https://www.business-humanrights.org/ja/%E6%9C%80%E6%96%B0%E3%83%8B%E3%83%A5%E3%83%BC%E3%82%B9/dahua-technology-did-not-respond-to-allegations-about-sale-of-surveillance-technology-in-latin-america-questions-from-cso-on-human-rights/

Dratwa, J. (2014). *Ethics of Security and Surveillance Technologies* (28; Opinion). European Group on Ethics (EGE). https://grundrechte.ch/2014/opinion_28_securityandsurveillancetechnologies.pdf

Dwoskin, E. (2018, May 22). Amazon Is Selling Facial Recognition to Law Enforcement—For a Fistful of Dollars. Washington Post. https://www.washingtonpost.com/news/the-switch/wp/2018/05/22/amazon-is-selling-facial-recognition-to-law-enforcement-for-a-fistful-of-dollars/

Ellis, R. E. (2022). El Avance Digital de China en América Latina. *Revista Seguridad y Poder Terrestre*, *1*(1), 15–39. https://doi.org/10.56221/spt.v1i1.5

European Union. (n.d.). *Biometric Data—European Commission*. Retrieved July 12, 2024, from https://home-affairs.ec.europa.eu/networks/european-migration-network-emn/emn-asylum-and-migration-glossary/glossary/biometric-data_en

Faraco Cebrian, F., do Amaral Prudente, G., Guedes, M. S., Ferreira da Silva, M. C., Duarte, M. L., & Guimarães Moraes, T. (2024). *Biometría e reconhecimiento facial* (2; Radar Tecnologico). Autoridade Nacional de Proteção de Dados. https://www.gov.br/anpd/pt-br/documentos-e-publicacoes/radar-tecnologico-biometria-anpd-1.pdf

France 24. (2020, December 10). *Economía—Cuba renuncia a su participación mayoritaria en empresas mixtas para fomentar la inversión*. France 24. https://www.france24.com/es/programas/econom%C3%ADa/20201210-cuba-renuncia-a-su-participaci%C3%B3n-mayoritaria-en-empresas-mixtas-para-fomentar-la-inversi%C3%B3n

Frescura Toloza, D. E. (2019). Debates públicos en torno a la creación del Sistema Federal de Identificación Biométrica (SIBIOS): Tensiones entre seguridad y privacidad. *XIII Jornadas de Sociología. Facultad de Ciencias Sociales, Universidad de Buenos Aires*. https://cdsa.aacademica.org/000-023/410.pdf

Gates, K. A. (2011). *Our Biometric Future: Facial Recognition Technology and the Culture of Surveillance*. NYU Press.

Gonzalez Omerod, A. (2022, April 22). *How AI Reinforces Racism in Brazil*. Rest of World. https://restofworld.org/2022/how-ai-reinforces-racism-in-brazil/

Government of Coahuila, Mexico. (2020). *Coahuila cuenta con el mejor sistema de video vigilancia en seguridad: MARS*. https://coahuila.gob.mx/noticias/index/coahuila-cuenta-con-el-mejor-sistema-de-video-vigilancia-en-seguridad-mars-23-06-20

Government of Panama. (2018, November 21). *Primer Centro de Operaciones de Seguridad y Emergencias C2 resultado de la cooperación entre Panamá y China*. Ministerio de Relaciones Exteriores. https://mire.gob.pa/primer-centro-de-operaciones-de-seguridad-y-emergencias-c2-resultado-de-la-cooperacion-entre-panama-y-china/

Haerpfer, C., Inglehart, R., Moreno, A., Welzel, C., Kizilova, K., Diez-Medrano, J., Lagos, M., Norris, P., Ponarin, E., & Puranen, B. (2022). *World Values Survey Wave 7 (2017–2022) Cross-National Data-Set* (Version 4.0.0) [Dataset]. [object Object]. https://doi.org/10.14281/18241.18

Hannig Núñez, S., & Ichihara, M. (2024). TikTokが安全保障に与える影響とTikTok規制の現在. 情報法制研究. https://www.yuhikaku.co.jp/books/detail/9784641498020

IDEMIA. (2021). *Brazil Tests the World's First Facial Recognition Shuttle Service*. https://www.idemia.com/pdf-export.php?post_id=7043

Infobae América (Director). (2013, June 27). *Infobae: Entrevista a Julian Assange* [Video recording]. https://www.youtube.com/watch?v=If7MbOvuEbg

Johnson, K. (2023). FBI Agents Are Using Face Recognition without Proper Training. WIRED. https://www.wired.com/story/fbi-agents-face-recognition-without-proper-training/

Königs, P. (2022). Government Surveillance, Privacy, and Legitimacy. *Philosophy & Technology*, *35*(1), 8. https://doi.org/10.1007/s13347-022-00503-9

Kostka, G., Steinacker, L., & Meckel, M. (2021). Between Security and Convenience: Facial Recognition Technology in the Eyes of Citizens in China, Germany, the United Kingdom, and the United States. *Public Understanding of Science*, *30*(6), 671–690. https://doi.org/10.1177/09636625211001555

Lagos, M. (2023). *Informe 2023 LA RECESIÓN DEMOCRÁTICA DE AMÉRICA LATINA*. Latinobarómetro. https://www.latinobarometro.org/lat.jsp

Lyon, D. (2001). Facing the Future: Seeking Ethics for Everyday Surveillance. *Ethics and Information Technology*, *3*(3), 171–180. https://doi.org/10.1023/A:1012227629496

M. L., R. (2023, March 16). *Interview with a Former High-Level Politician, Colombia.* [Online].

Macnish, K. (2017). *The Ethics of Surveillance: An Introduction*. Routledge.

Majerowicz, E., & Carvalho, M. H. de. (2024). China's Expansion into Brazilian Digital Surveillance Markets. *The Information Society*, *40*(2), 168–185. https://www.tandfonline.com/doi/abs/10.1080/01972243.2024.2315880

Metrô define empresa que vai fornecer sistema de reconhecimento facial por R$ 58,6 milhões. (2019, October 19). Diário do Transporte. https://diariodotransporte.com.br/2019/10/19/metro-define-empresa-que-vai-fornecer-sistema-de-reconhecimento-facial-por-r-586-milhoes/

Morena-Álvarez, C., Vila-Seoane, M., Morena-Álvarez, C., & Vila-Seoane, M. (2023). La cooperación entre Ecuador y China en tecnologías de seguridad: El caso del ECU 911. *URVIO Revista Latinoamericana de Estudios de Seguridad*, *36*, 85–103. https://doi.org/10.17141/urvio.36.2023.5847

Mozur, P., Kessel, J. M., & Chan, M. (2019, April 24). Made in China, Exported to the World: The Surveillance State. The New York Times. https://www.nytimes.com/2019/04/24/technology/ecuador-surveillance-cameras-police-government.html

Nuevo sistema de Videovigilancia Urbana y Monitoreo en la ciudad de Pico Truncado. (2020). Revista Innovación Seguridad. https://revistainnovacion.com/nota/10889/nuevo_sistema_de_videovigilancia_urbana_y_monitoreo_en_la_ciudad_de_pico_truncado/

Pelcastre, J. (2019, October 22). *China exporta modelo de control ciudadano a Bolivia.* Diálogo Américas. https://dialogo-americas.com/es/articles/china-exporta-modelo-de-control-ciudadano-a-bolivia/

Pisanu, G., & Arroyo, V. (2021). *Surveillance Tech in Latin America* (pp. 1–53). Access Now. https://www.accessnow.org/wp-content/uploads/2021/08/Surveillance-Tech-Latam-Report.pdf

Project | china.aiddata.org. (n.d.). Aid Data. Retrieved July 23, 2024, from https://china.aiddata.org/projects/39281/

Proveedores identificados. (n.d.). Reconocimiento Facial. Retrieved July 6, 2024, from https://estudio.reconocimientofacial.info/proveedores-identificados/

Querido, L. (2021). *El modelo liberal cubano y su influencia en América latina* (Vol. II, 1st ed.). Editorial Dunken. https://repositorio.4metrica.org/bitstream/handle/001/78/1.%20El%20modelo%20liberal%20cubano%20y%20su%20influencia%20en%20Am%c3%a9rica%20latina%20%28Vol.%20II%29.pdf?sequence=1&isAllowed=y

Ramiro, A., & Cruz, L. (2023b). The Grey-Zones of Public-Private Surveillance: Policy Tendencies of Facial Recognition for Public Security in Brazilian Cities. *Internet Policy Review*, *12*(1). https://doi.org/10.14763/2023.1.1705

Response from IDEMIA to Allegations about Sale of Surveillance Technology in Latin America. (2022). Business & Human Rights Resource Centre. https://www.business-humanrights.org/ja/%E6%9C%80%E6%96%B0%E3%83%8B%E3%83%A5%E3%83%BC%E3%82%B9/response-from-idemia-to-allegations-about-sale-of-surveillance-technology-in-latin-america/

Rodriguez, K. (2012, January 10). *Biometrics in Argentina: Mass Surveillance as a State Policy.* Electronic Frontier Foundation. https://www.eff.org/deeplinks/2012/01/biometrics-argentina-mass-surveillance-state-policy

Smith, M., & Miller, S. (2022). The Ethical Application of Biometric Facial Recognition Technology. *AI & Society*, *37*(1), 167–175. https://doi.org/10.1007/s00146-021-01199-9

Souza, M. R. O., & Zanatta, R. A. F. (2021). The Problem of Automated Facial Recognition Technologies in Brazil: Social Countermovements and the New Frontiers of Fundamental Rights. *Latin American Human Rights Studies*, *1*. https://revistas.ufg.br/lahrs/article/view/69423

Tigre lanzó NeoCenter, un sistema de vanguardia en reconocimiento facial. (2019, May 15). https://www.minutouno.com/politica/tigre/lanzo-neocenter-un-sistema-vanguardia-reconocimiento-facial-n5031770

U. S. Mission Chile. (2021, July 19). *EE.UU. adopta medidas contra entidades que facilitan abusos de derechos humanos en China.* U.S. Embassy in Chile. https://cl.usembassy.gov/es/estados-unidos-pretende-bloquear-la-vigilancia-del-gobierno-chino/

US Blacklists China Organisations over Xinjiang "Uighur abuse." (2019, October 7). BBC. https://www.bbc.com/news/world-us-canada-49968126

Venturini, J., & Garay, V. (2021). *Reconocimiento facial en América Latina: Tendencias en la implementación de una tecnología perversa.* Al Sur. https://www.alsur.lat/sites/default/files/2021-11/ALSUR_Reconocimiento_facial_en_Latam_ES.pdf

Wilson, J. M. (2024, July 5). *Misión de diputados viaja al "Silicon Valley" chino para aprender de inteligencia artificial.* La Tercera. https://www.latercera.com/politica/noticia/mision-de-diputados-viaja-al-silicon-valley-chino-para-aprender-de-inteligencia-artificial/PFRQHACXF5DW3BRAODRUJZQUX4/

Zureik, E., & Hindle, K. (2004). Governance, Security and Technology: The Case of Biometrics. *Studies in Political Economy*, *73*(1), 113–137. https://doi.org/10.1080/19187033.2004.11675154

9 5G Networks in Latin America

Caught in a Global Geopolitical Conflict between China and the World?

Despite expectations and media attention, 5G deployment in Latin America is taking longer than anticipated. While Chile and Brazil were pioneers in launching their networks and are looking ahead to a second and third round of spectrum auctions,[1] other countries have only recently achieved widespread 4G coverage.[2] The unevenness makes it difficult to generalize a "regional" trend or decision-making rationale.[3] This has not prevented researchers from analyzing the broader geopolitical debate in this regional and cultural context,[4] since 5G network diffusion symbolizes Beijing and Washington's technological and arguably ideological competition globally.[5]

The evidence I present in this chapter shows a complex story reinforcing the idea of power dynamics affecting the outcome. However, this is not the result of direct competition between private actors from the US and China. Process tracing analysis of the Brazilian, Chilean, Argentinian, and Colombian cases initially shows that American companies are not even present in 5G spectrum auctions in the region; Huawei has found two significant schemes to enter the market: price competition and dealing directly with governments and institutions—while Ericsson and Nokia are long-standing partners for some leading telecommunication companies that are present from Mexico to the Antarctic territories. The US has followed key states' decision-making from the beginning and has challenged Chinese dominance even if it doesn't directly incentivize US-based companies to enter the market. For example, during the early stages of 5G diffusion, Washington officials lobbied Brazil and Chile to exclude Chinese providers from their processes—without much success in changing these countries' guidelines.[6] Huawei, a company with a robust commercial presence in Latin America, launched an intense public diplomacy campaign amidst the 5G deployment process. This included deployment tests focused on South America[7] and training programs in collaboration with regional governments.[8]

Supported by interviews,[9] primary, and secondary sources, I also argue that even though system preferences and partnerships of service providers are not always openly disclosed—Huawei, Nokia, or Ericsson[10] do not bid directly—this does not mean that governments are unaware of the *most likely* source of technology. Documents and accounts show authorities are aware of the gravity of the international tensions between the US and China. They resort to open sources,[11] past auctions, press articles, internal discussions, and their built-up knowledge to forecast

DOI: 10.4324/9781003489450-11

how service providers will engage with technology companies. Moreover, in the more stable democracies in the region, there is an expressed underlying pressure to answer to domestic public opinion, which is influenced and mirrored in the media. To capture this discourse, I conducted a text analysis on a sample of mainstream media outlets' publications in three countries: Argentina, Chile, and Colombia.[12] The results of this analysis—later introduced in this chapter—indicate that the media in each country has a different approach to the process. However, outlets generally emphasize the 5G securitization debate, and "Huawei" is broadly mentioned over other competitors.

This chapter briefly describes the process, the main actors, and the mechanisms through which 5G public tenders have developed in Latin America. It evaluates economic, geopolitical, and value-based conditions that interact with the outcomes in each country. My conclusions follow the results of both types of analysis and create a picture of the nature of the region's outputs. Most importantly, the 5G diffusion trend in Latin America reflects a critical idea: the US has only limited influence in states' decision-making, and Chinese companies have not taken the role they were expected—as shown in both literature and media coverage—in terms of technological leadership in the region. However, telecommunication providers might be the entities most affected or influenced by international trends, such as the ban on Huawei in key territories such as the US, Australia, and some European countries. This is because *telecom* operators sometimes participate from both markets that have restricted Huawei as a provider and those that are open to any offers, and prefer to avoid the costs of partial bans in their supply chains.

The Road of 5G Development in Latin America

Introducing mobile networks was a groundbreaking development for emerging nations and the whole of Latin America. Home access to the internet, restricted to the ownership of hardware and monthly subscriptions to a network provider, is prohibitive for many families in the region, which causes inequality in the access to information and services that had increasingly migrated through digitalization,[13] something that became evident during the COVID-19 pandemic when online classes and remote work became a norm in many countries. In comparison, mobile networks tend to be cheaper and more accessible, and—though they don't secure a stable connection—they are an equalizing factor in societal relations. In other words, mobile telecommunications have become a pillar of Latin American societies.[14]

This is mirrored in regional connectivity statistics. While countries in South America have connectivity levels among the highest in the world, many of those connections are explained by the widespread use of mobile connections.[15] Chile, for example, exhibits a net internet penetration of 97.2%.[16] However, only 67% of households are estimated to have a fixed internet contract,[17] while the penetration of mobile connections sits at 105%[18]; in other words, there are 105 mobile connections for 100 individuals.

Despite the need for and importance of rapid and stable mobile connections, Latin American countries have—comparatively—been slower at deploying 5G networks than other developing regions. In this sense, there is still debate on what companies will dominate the Radio Access Network (RAN) in Latin America.

Commercial trials have been deployed since around 2018. Still, the actual government-led spectrum bids have encountered a series of delays related to factors that range from geographically challenging terrain to unstable and changing politics. Brazil, Uruguay, and Puerto Rico were the first to launch commercial 5G in the region. Chile was the first to officially announce a 5G spectrum public auction,[19] which is now advancing to its second phase. Meanwhile, Central American countries are still behind even in 4G connectivity and net internet penetration, with an average of 61.9%, and some countries—such as Honduras—with under 40% of net internet access.[20] Mexico appears to have a unique position as with other technologies, mainly due to its economic ties with the US.

In this context, the 5G debate and subsequent decision-making efforts have taken different regional approaches. Upon consultation via interviews, I identify two baselines of opinion regarding this phenomenon among political actors. The first line is the need to enhance the importance of privacy, which is only security and value-related factors in the decision-making process. That is both from a deontological perspective and a more technical one, mentioning aspects such as reputation, preference for "like-minded" systems, and the pursuit of privacy integrity.[21] Two organizations have guided regional leaders in what is considered a "standard." One is the OAS, through its cybersecurity framework,[22] but most importantly, through the Inter-American Telecommunications Commission (CITEL), which consolidates public and private proposals for using spectra.[23] The second one is the International Telecommunication Union (ITU), which aims at establishing norms that allow for global connectivity and coordination[24] and allocates portions of the radio-electric spectrum and services in frequencies by dividing the world into regions, inserting Latin America within Region 2 of this categorization.[25]

However, since the early 2010s, there has been a visible change in the position of countries from an overall attempt at establishing universal standards that respond to the notion of international liberal order—to a divided scenario in which "like-minded" countries aim at pushing changes or maintaining status quo in these universal conventions. Public officials identify these blocs as a "China, Russia, some Arab countries, and BRICS" side and a "Western" bloc including Australia—the first to implement restrictions against Chinese companies—New Zealand, the US, or Canada.[26]

Several conditions related to the Chinese market and CCP intervention in their economy[27] raise questions about the reliability of PRC-based telecommunication companies operating overseas.[28] Another example is the National Intelligence Law of China, which was approved in 2017. Article 7 of the law specifies that "organizations and citizens are obliged to support, assist, and cooperate with [PRC] intelligence organs." Article 14 authorizes the intelligence organs to demand assistance from institutions, organizations, and citizens—a primary concern for the US and other countries as Chinese companies expand globally.[29] At the same time, the PRC has long restricted foreign technologies and digital services to operate domestically, limiting the extensive Chinese market to international competitors. Because of this and other related conditions—such as the proven link between Huawei supplying Iranian companies with American technology[30]—there is a pending argument on whether Huawei and other Chinese companies play "by the book" following all domestic and international regulations concerning Private Data and security in Latin America.[31]

The same logic applies to Russia's interests, as it has increasingly lobbied to change international information and technology norms.[32] Although not a prominent player in innovation, I recognize an ongoing debate on how to engage with these disruptive actors. The 5G network technology transfer case provides just one outcome of this "bloc" skepticism.

The second widely spread opinion is a *pragmatic* recognition of situational conditions and limitations that states face. For example, in some cases, there may not be enough providers competing in an auction or public induct, so the decision-making apparatus was able to come up with the selection that best matches the state's needs. In other cases, reputation or concerns over the selected system are overshadowed by intrinsic elements such as political or economic dependency–corruption, lack of consistency, opportunism from external and internal actors, ignorance of the importance of private data, or the relegation of responsibility between bodies.[33]

The influence of internal conditions in both scenarios is consistent with neoclassical realism and Robert Putman's two-level games, as the outcomes of these technology public tenders reflect a state's foreign policy and often affect the target country's relationship with the providers. Evidence—including the cases I review in this book's chapters—seems to point to the fact that although private companies might not be directly linked with their home country's government, except for the case of China in which there is a legal basis to assume this. In practice, decision-makers do link the origin of a provider with a particular expected culture, compliance, and reputation. However, there is an ongoing debate over whether this notion and understanding affect the final policy output of each state. This is particularly important considering policymakers' focus on the tradition of *neutrality* in the Latin American region,[34] which might interfere with any nation's aims to find direct doors to influence technology diffusion.

The Quest for an Ideal Neutrality and Non-discrimination

Despite differences, the general norm for regional institutions is to abide by the principles of neutrality and non-discrimination when determining which companies can participate in a public tender process, as regulators and IT-related bodies tend to be more technical in their assessment.[35] Non-discrimination in public tenders aims to avoid differentiated treatment of providers participating in public tender processes.[36] As a result, screening potential providers is limited to specific conditions needed to ensure reliability—such as a company's track record or if it has been involved in notable corruption cases—but that seldom include security considerations as it happens in the US. The screening is, however, sometimes included after the companies bid for a particular project. On the other hand, network neutrality principles—related to the nature of the service—in the Latin American region[37] have faced 5G as a challenge. Still, laws have been flexible enough to enable governments to face the existing concerns over telecommunications and to allow for competition in 5G networks.[38]

Regarding the 5G network diffusion process in Latin America, the conditions for network auctions have varied depending on the country. President Jair Bolsonaro initially threatened to ban Chinese providers from the process due to security concerns in Brazil. Despite this narrative and US pressure,[39] the process was open to participants without restrictions. It could have also been influenced by China's overwhelming economic grip on Brasilia and its well-established political ties.[40] In Chile, 5G was primarily related to a technical discourse ruled by a strict policy regarding neutrality in investments, neutrality on the internet, and technological neutrality, which act as an umbrella for the decisions at a government level. Public officials tend to agree that public opinion does pressure the process outcome, which should focus on quality and whether the base structure, materials, and rods are of acceptable quality.[41] There is a political debate in Colombia regarding the deployment of the 5G network. Officials see neutrality and non-discrimination as keystones for their decision-making process based on legislation and consider it a value-based process, also considering the influence of international organizations. The Colombian law on "Information Society, Management of Information Technologies and Communications" that establishes the National Spectrum Agency states that *technological neutrality* is an ethical principle and that:

> The state will guarantee the **free adoption** of technologies, taking on account recommendations, concepts, and normative principles from **International Organizations** which are relevant to the matter, which allow the effective facilitation of services, contents, and, applications that use information and communication technologies to guarantee free and **fair competition** and to ensure that their adoption balances development with environmental sustainability.[42]

Colombia perceives neutrality as a value closely related to economic and political freedom. However, opinions within the decision-making system vary when facing a more political and security-related debate, such as the 5G network deployment. The same is true for the reliance on international norms that are now being challenged. In other words, there is a recognition of potential pressure and changes in the international system, but this does not always mirror the output of public technology auctions.

A case that stands out in comparison is Costa Rica's 5G regulation process. In August 2023, Costa Rican President Rodrigo Chaves approved a decree on deploying 5G mobile networks. In short, the decree banned tech companies from countries that have not endorsed the Budapest Convention on Cybercrime, which included China, South Korea, Russia, and Brazil.[43] For many, the decision was an excuse to avoid the direct banning of Huawei but an effective limitation for bidders. Nevertheless, Costa Rican Courts suspended this decree "until this authority has greater evidence to definitively resolve the request for a precautionary measure," as requested by the Internal Workers' Front (FIT) within the Costa Rican Institute of Electricity (ICE).[44]

Overall, Latin American states have avoided discriminating against companies in their 5G network auctions. The overall justification for this is market and political principles. However, American and Chinese lobby activities can be connected to some of these decisions and outcomes, as in the cases of Brazil and Costa Rica. Understanding the aim for neutrality is essential, but even with this principle—as seen in other instances, *Government-to-Government* agreements are constantly taking place, reflecting the influence of primary and middle regional powers. There are essential difficulties measuring the overall presence of these companies since—in addition to institutional projects—they also engage in localized private endeavors that can directly compete without a top-down government policy.

Moreover, I identified a general agreement among crucial actors that there needs to be more concern around the security issues related to data sharing and data governance across regional states. This applies to Chinese providers and any company that enters the market.[45] The consequence is incompatibility with more robust standards and vulnerability for the local population. In this scenario, there is a shared perception that European companies might provide relatively safer service due to the restrictions they face in their country-of-origin jurisdictions.

The Long-Term Players in Telecommunications

The process of 5G offers in Latin America is mainly carried out by well-established mobile service providers.[46] This makes it feasible to predict which technology companies will likely be involved in each 5G process. "América Móvil," a Mexican enterprise that operates through its subsidiary "Claro," has a wide presence in Argentina, Brazil, Chile, Colombia, Costa Rica, Ecuador, El Salvador, Guatemala, Honduras, Nicaragua, Panama, Paraguay, Peru, Puerto Rico, Dominican Republic, Uruguay, the US, and Spain.[47] Telefónica, which operates as Movistar in Latin America, is a Spanish company with a presence in Argentina, Brazil, Chile, Colombia, Ecuador, Peru, Mexico, Uruguay, Spain, and Venezuela.[48] Tigo, a Colombian-European company, is in Colombia, Bolivia, Costa Rica, El Salvador, Guatemala, Honduras, Nicaragua, Panamá, and Paraguay. Finally, Entel, a Chilean company, is present in Chile and Peru. WOM, a company under the British Novator Partners, operates in Colombia and Chile and is closely related to China. Between 2015 and 2019, the China Development Bank provided WOM with $415 million in loans in exchange for granting Huawei preferred network partnerships.[49] In contrast, the US is primarily present in Central and North America, with scarce direct participation in the South American market.

As for technology manufacturers, there is also a limited offer that concentrates on companies from Scandinavia and China, with US. providers playing only a minor or insignificant role in the race for network deployment. This is interesting because Washington has been outspoken in warning Latin American countries about the presence of Beijing-linked companies, but there isn't a direct economic interest behind these actions. In other words, American providers do not directly benefit from any hypothetical measure against China since they are absent in the competition to earn those contracts. I find this necessary because it mirrors the overarching debate of this book: technology is a central component in both China's extensive international influence and the increasing skepticism from the US toward China, a phenomenon that is not exclusive to these two countries but part of a broader debate on security, privacy, and threat perception in the context of hybrid warfare and cognitive warfare concerns.

Understanding private and public dynamics in disseminating 5G technology is essential. Nokia (Finland) and Ericsson (Sweden) have an extended presence across most of the region's countries. Both were popular cellphone sellers in the early 2000s, and their strategy in Latin America shifted to a more technical one as they saw their presence diminished by new competitors. In that sense, they negotiated with carriers and have gained a position as default manufacturers for the technology behind the service. In a presentation by the Chilean Telecommunications Subsecretary (Subtel) to the Senate, they emphasized that in the year 2014, the 4G network auction was decided in favor of Entel along with Ericsson, Claro along with Nokia, Telefónica with both Nokia and Huawei and WOM that worked with Huawei[50] following its partnership. This pattern of business collaboration has been stable over time and across the region. Nevertheless, since the US increased its pressure to reduce Chinese presence in its market, some of these companies' conditions have seemingly changed. Experts agree that, despite countries not imposing restrictions on Huawei as a manufacturer of 5G infrastructure, service providers might decide to limit their economic partnerships with Chinese entities.[51]

This was the case for Telefónica. The Spanish company has been working with Huawei for years and has provided network services to countries in Europe and Latin America. Because of this partnership, the Chinese manufacturers manage many 4G networks, and one could assume that the company would maintain this strategy during the 5G deployment process. However, since 2019, Telefónica has reduced Huawei's presence as a provider, allegedly as a technical decision,[52] but clearly as the consequence of political events worldwide,[53] as other European operators also leaned toward alternatives to Chinese companies due to potential restrictions.[54] The process after the initial announcement has encountered contradictory statements. By the end of 2023, Telefónica was finalizing a tender to replace Huawei as a provider,[55] removing the Chinese company from its significant markets,[56] while considering potential legal actions in Germany,[57] being sanctioned by the Securities and Exchange Commission (SEC) for their business with Huawei in Venezuela,[58] and participating in 5G network adjudication processes in the Latin American region, still working with Chinese providers in other areas. Nokia was the company that replaced Huawei in Telefónica's service-providing schemes.

Huawei has exponentially grown by establishing alliances with governments, smaller suppliers, state enterprises,[59] and direct investments. It was also among the companies that invested in commercial trials across Latin America. By 2021, it had launched deployment tests and demonstrations in Argentina, Brazil, Bolivia, Chile, Colombia, Ecuador, and Peru.[60] The company has also invested in promoting its brand through thorough marketing, training,[61] and collaboration campaigns with private and public entities.

It is important to note that there are different ways to measure these providers' market share. Some focus on sales and net income, while others focus on radio access networks or RAN share. Generally, Huawei performed well internationally, with an estimated share of over 30% of the global market[62] when the Chinese market is included, and a significantly reduced share when excluding mainland China. In Europe, a Strand Consult study calculates that Chinese vendors held up to 41% of the 5G RAN market in 2022, mainly due to the importance of the German economy,[63] but this is a reduction compared to the Chinese share of 4G Ran, which was 51%, showing a significant decrease.[64] The number could decrease due to changes in European frameworks. In particular, the EU 5G toolbox is expected to cause a de facto removal of major Chinese telecom providers from regional networks.[65]

Despite the situation in the European and American markets, in most metrics, the competing actors are Huawei, Nokia, and Ericsson, with Samsung and ZTE as alternatives. The last two items have very little presence in Latin America.

Economic Incentives and 5G Diffusion

I withdraw that—even among key actors—there is a general belief that Chinese providers compete in price and not in providing quality or ensuring the security of their products. The reputation of Chinese providers is still hindered by the widespread notion of a cheap offer that will only comply with minimum requirements.[66] However, since some Latin American countries need to provide their citizens with accessible connectivity, a lower offer can be the wisest way to ensure connectivity and the social benefits of—in this case—mobile connections. In these cases, Chinese companies—and their mobile carrier partners—could hold the advantage.

Huawei enjoyed overwhelming advantages in the Latin American market in 2014. First, it was due to its lower costs, and second, it was due to its "flexible" customer experience.[67] Huawei was—unsurprisingly—expected to be the primary winner in 5G spectrum auctions in Latin America, but this has not been the case across the region. In Table 9.1, I summarize the current state of 5G network deployment in Latin America.

The results of the Uruguayan, Brazilian, and Chilean 5G public tenders show that states have shared the spectrum among several providers. Even though some winners work with Chinese companies, this has not matched the expectations set up before Beijing became directly confronted with its policies regarding technology and overseas management of IT companies. Telefonica (Vivo), a Brazilian 5G Network system provider, launched a tender in that market in 2022 to effectively replace Huawei, a decision that directly benefited Nokia.[68] In Uruguay, Claro (América Móvil) and Movistar (Telefónica) won the 5G network public tender, further consolidating the idea of reducing Chinese participation in the region. In Chile, the spectrum was assigned to practically all bidders: Entel, Wom, and Movistar (Telefonica) won the initial process, and Claro (América Móvil) was

Table 9.1 Results and most likely providers of several 5G deployment processes in Latin America, portraying market share and the interactions between companies. Main source: Bernal (2024).

Network deployment in Latin America (2020–2024)				
Country	*Date*	*Round*	*Assigned to*	*Partnership with (most likely/proven)*
Chile	2021	First	Entel	Ericsson
			Movistar (Telefónica)	Nokia (and Huawei)
			Claro	Nokia
			WOM	Huawei
Brazil	2021	First	Algar Telecom	Huawei, Nokia and Ericsson
			Claro Brasil	Nokia
			TIM Brasil	Huawei (and Ericsson)
			Telefónica Brasil (Vivo)	Nokia and Huawei
Mexico	2022	First	AT&T	Samsung, Ericsson and Nokia
			Radiomovil Dipsa (Telcel	Nokia and Ericsson
			Telefónica Móviles México (Movistar)	Nokia (and Huawei)
Argentina	2023	First	Personal (Telecom)	Nokia (and Huawei)
			Movistar (Telefónica)	Nokia (and Huawei)
			Claro	Nokia (and Huawei)
Colombia	2024	First	Claro	Nokia
			Movistar (Telefónica)	Nokia
			Tigo	Nokia
Uruguay	2023	First	Movistar (Telefónica)	Nokia
			Claro	Nokia
República Dominicana	2021	First	Claro	Nokia
			Altice	Nokia
Guatemala	2023	First	Tigo	Nokia
			Claro	Nokia
			State-owned	N/A

authorized to utilize part of the spectrum later. This means that the most prominent technology providers are present in the country. Even though WOM openly collaborates with Huawei and has secured significant projects with them, the company has encountered financial challenges in recent years, filing for bankruptcy in the US in April 2024.

While decision-makers and public officials have worked to keep the conversation as technical as possible—reflected in both the interviews I conducted and my analysis of media behavior—there is consensus that public opinion, geopolitics, and diplomacy play a role in shaping the outcome of tenders for sensitive technologies. One public official even expressed that "if the price were the only factor [to decide an outcome], China would win many more projects [in Chile]."[69]

However, some other arguments reject the idea that Chinese providers are economically savvy for mobile companies to use when bidding for the 5G spectrum and that there aren't major costs in transitioning from Chinese providers of 4G to other 5G providers. Even more, there are economic risks for large-scale providers to use Chinese technology. First, there is a latent risk of supply chain disruption if sanctions are imposed or tensions escalate between critical countries. Second, there is a financial risk when a company is banned or even in access to credit and funding from organizations that have imposed restrictions.[70] Hence, mobile services are expected to avoid dealing with problematic providers. These factors can be costly for the country auctioning the 5G spectrum and the service provider that must secure continuous operations in the networks for which they have control and responsibility.

Geopolitical Conditions and 5G Network Diffusion

Telecommunications networks have become critical state infrastructure for security, stability, and social cohesion. The connectivity disruption within a nation is problematic, at least during emergencies and natural disasters. Thus, states emphasize how their telecommunications systems are laid out and how these networks help solve social inequalities such as access to services.[71]

As in other cases in this book, international trends and the actions of leading Western nations can impact Latin American states' behavior via lobbying or the pursuit of harmonizing systems with certain nations.[72] Geopolitical tensions also pressure some international institutions and organizations to limit the presence of certain companies—like those with PRC ties—as providers when they set up frameworks for sensitive technologies[73] in development projects. That is mainly to comply with security regulations in their base countries, to avoid criticism, or to avoid problems when they also operate in countries that limit providers for security reasons.[74]

Regarding telecommunications, a clear securitization discussion began in Australia in 2012, gained traction in the US around 2018, and later in the European Union. In this last case, risk assessment has been a central element of debate and a keystone of the EU response to this issue. One concern being pushed by the EU cybersecurity agency is "third-country interference on a supplier,"[75] something that immediately affects the operations of Chinese companies in the market since there is a basis to argue State control, participation, and influence over telecommunications is a real risk concerning PRC providers.[76]

In Latin America, tensions between the US and China are part of some political discourses. For example, in a Senatorial discussion in Chile in 2020 that included the presence of members of the executive power, academia, and civil society, one senator pointed out that the 5G network discussion was taking place "amidst a technological war" which involved "Nokia, Ericsson, Huawei, ZTE, and Samsung." A second policymaker stressed that the US and China are competing in the region to increase their influence over data and artificial intelligence, and a third policymaker confirmed that they were looking at the US and Europe for their standards in the public bidding process,[77] which means Chile could mirror some of those decisions.

In Colombia, circumstantial conditions have changed the perception of technology-related threats, which could affect policy output. Left-wing president

Gustavo Petro is expected to be much closer to China and PRC companies than its predecessor, Iván Duque, who was closer to the US.[78] On the other hand, I identify a shared discourse among politicians on the defense of certain principles when debating on the origin of technology. Value affinity among democracies, reputation, and national security issues are listed among the reasons to avoid choosing Huawei in particular, but as a reaction to the US' decision to blacklist the company.[79] Colombia is also openly influenced by what global and regional pioneers have been doing regarding their 5G deployment strategy. For example, an official document on the 5G deployment process mentions the United Kingdom—as for the diffusion of international standards—Spain, Brazil, South Korea, Japan, China, and Chile as examples of countries that have implemented successful strategies and that have developed this technology[80]. It also mentions the European Union, the UK, Spain, the US, Germany, and Brazil, as well as Chile and Mexico, as models to follow in terms of 5G public tenders.[81]

The geopolitical discourse is widespread, clearly stated, and impacts the region's decision-making. This is either in the private sector, as companies might prefer to change providers due to regulations in some of the countries they operate in, or in the public sector, as states can be influenced and pressured by domestic affairs and public opinion.

Public Discourse and 5G in Latin America

Given the consensus among public officials and experts regarding the significance of public opinion and its influence on decision-making, examining the nature and status of this discourse can be a valuable tool for understanding the circumstances and context in which this phenomenon occurs.

Upon reviewing the narrative that mainstream newspapers and digital media[82] from Argentina, Chile, and Colombia between 2019 and 2022 regarding "5G," I highlight that China and the US—in the context of a trade war—are at the center of the security debate even though US companies do not provide 5G services to these countries at a national level. It is evident that "national security" concepts indicate a recurring concern about the issue. This is something decision-makers mention, as public opinion inquires the public sector to address this international debate, even if public officials prefer to keep the discussion as technical as possible. Figure 9.1 represents the most common words—ignoring stop words in Spanish—found in the reviewed articles.

The word "Huawei" is the fourth most frequently used word by these media outlets. It is strongly associated with China and national security-related terms, as well as words related to threat and conflict. Terms related to cybersecurity and data protection are slightly related to this. In Argentina, concepts related to security are much more present in media discussions than in Chile or Colombia. Journalists often analyze these issues, connecting Chinese influence in the country with the upscaling regional tensions between Washington and Beijing.

In comparison, none of the other companies even appear among the 50 most used words when analyzing the data pool. Therefore, the role of essential actors with more presence in the market, namely Nokia, Ericsson, and their origin countries, Finland

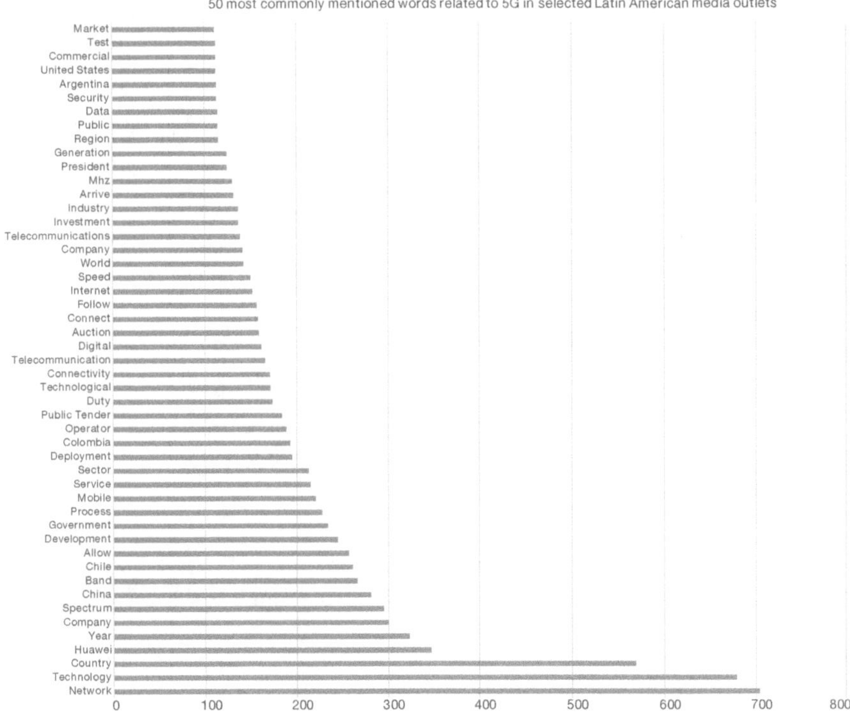

Figure 9.1 A list of the most recurrent words found across publications related to 5G deployment in Argentina, Chile, and Colombia. The author's work uses a KH coder to process the text data for analysis.

and Sweden, are minimized in the public discussion and generally clustered. This could be due to Huawei's aggressive media strategy in the region or the perceptions of these local actors. Moreover, even when I look for these companies specifically, they are not narratively connected to the geopolitical conflict.

A technical and economic discourse is also present, mentioning issues such as the price, offer, spectrum, operators, RAN, or network—followed by a value-based cluster of terms, including "quality," "development," and "connectivity," and in all countries, a connection can also be found with concerns over poverty and the social and economic effects of improving connectivity for citizens. In Chile, the technical debate is the most important, and the geopolitical issues are secondary. Journalists maintain close narrative coordination with the regulating institution regarding what words they use and how they refer to the public tender process. In other words, media helps the states highlight the positive outcomes of improving internet access across the population, consistent with public officials' narratives and overall objectives stated in documents. In Figure 9.2, I present a general picture of how relevant keywords in the debate on 5G implementation in Latin America relate and interact. These results also show association and co-occurrence clusters. As with

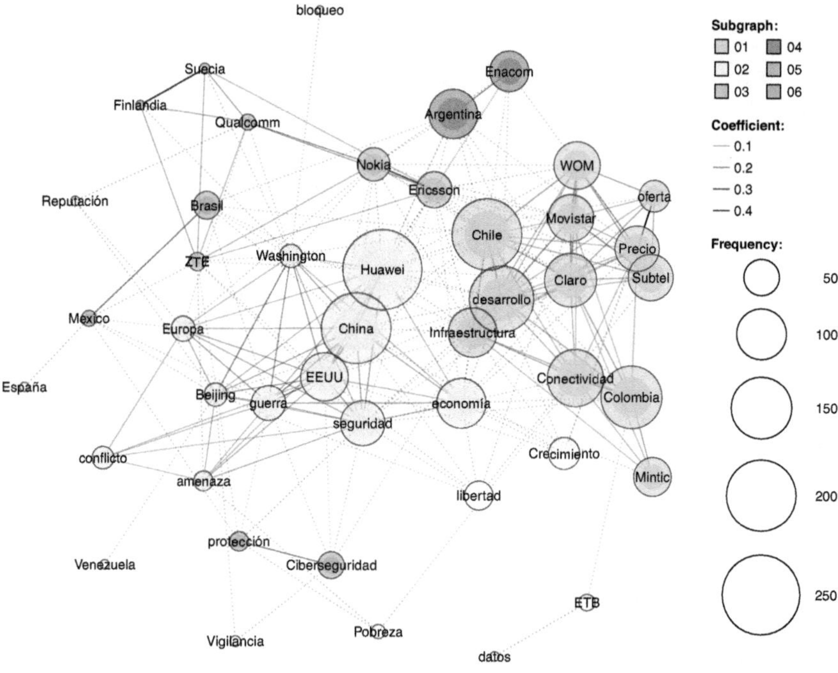

Figure 9.2 Coded terms related to the 5G deployment debate on the media in Argentina, Chile, and Colombia. The figure demonstrates the specific terms' recurrence, association, and co-occurrence. The author's work uses a KH coder to process the text data for analysis.

the previous table, these results come from an analysis of significant media outlets in Argentina, Chile, and Colombia.

Media outlets in Latin America have boosted exposition for Huawei and highlighted the escalating geopolitical tensions between Beijing and Western developed nations—mainly the US. In that sense, a narrative over reputation that often overlooks other companies has been spread throughout the region. Indifferently to theoretical notions on a second or new Cold War, public opinion is continuously exposed to an open debate on security, the rising importance of private data, and values related to political and economic systems that contrast with the globalization narratives that dominated the early 2000s. Media outlets also emphasize the technical aspects and benefits of 5G diffusion, focusing on economic development and connectivity for countries currently deploying critical telecommunications infrastructure.

I argue that all these clusters in publications are expected to impact or exercise pressure on decision-makers indirectly, as they consider both public opinion and international trends that are not always technical or purely economically driven. There is also proof of the impact of the "securitization" rhetoric across the region. In other words, what is happening with Huawei and other Chinese providers in

Australia, Europe, and the US reaches the Latin American media space and audience directly, also shaping some discourse around potential threats to telecommunications infrastructure and the idea of alignment in a new—and still debatable—form of Cold War or ideological division between pro and anti-liberal forces.

Conclusions and Theory

5G network diffusion in Latin America answers to domestic and international dynamics. On the one hand, states are incentivized to improve connectivity for citizens to boost development and service providing. Improved connectivity is also related to economic benefits, which is attractive for stakeholders involved in the process and for users who gain access to affordable internet and digital societal interactions.

On the other hand, decision-makers in Latin America mirror international regulation, relate to regional institutions, and use other neighboring countries as benchmarks for their 5G network deployment strategies. Both public officials and key private players, such as mobile providers, are aware of the widespread arguments against Chinese entities at several levels, which impact some parts of the technology diffusion procedure.

This reinforces my argument that we should resort to Robert Putnam's "two-level games" approach to explaining policy[83] when within-state debates and decision-making mirror the international scenario[84] and overarching norms crucial to the telecommunications framework. States in Latin America have had the freedom to address this challenge using their own technical and political tools, leading to noticeable procedural differences across the region. This further reinforces the idea of neoclassical realism as a framework to explain current technological diffusion mechanisms in international relations.[85]

Reputation is a hindrance factor for Beijing and a source of debate in the media, but on a scale different from Western developed countries. Huawei and ZTE are perceived as directly related to the PRC, and this deters actions at some stages of the deployment process. I identify a strong connection between private actors and reducing Chinese-related RAN. In other words, companies prefer to minimize risks and costs—of potential external sanctions or regulation—and limit their contracts with PRC providers. This is exemplified by the continuous processes to replace Huawei as a provider that Telefónica—a significant competitor in Latin America—has kickstarted in critical markets such as Brazil and Spain. The primary beneficiaries of Huawei's indirect restrictions have been Nokia and Ericsson, while Samsung has not yet directly entered the market.

Huawei's lower comparative share of 5G RAN in Latin America, compared to the previous 4G networks, cannot only be explained by private sector decisions and price competition since Chinese companies tend to offer lower costs and more flexible customer service. The region's norm has long been network neutrality and avoiding discrimination of bidding companies in public auctions, which is something that technical professionals and public officials vigorously defend. However, it seems that the combination of elements, including the increase in awareness among critical actors, pressures directed through public opinion channels, the increasing securitization debate impacting companies and institutions, the increased concerns over critical infrastructure, and the potential risk assessment both in terms of economic and social outputs has had an impact on the configuration of 5G diffusion in Latin America. Values and norms are sometimes mentioned as potential direct factors, such as in the case of Costa Rica and its screening framework for 5G. In other cases, like Chile, since decisions are strictly taken on a point-based system and the applicants are almost exclusively mobile operators, some deny the actual impact of shared values outside international norms and geopolitical issues on outcomes. In these cases, the influence takes effect through the commercial ties of long-term operators that might restrict Huawei to avoid economic and political risks.

The results show how the current ideological division between Western nations and challengers affects worldwide telecommunications infrastructure diffusion and innovation. The comparisons between 4G RAN and 5G share among technology providers further emphasize changes in decision-making, private strategy, and the interactions between states and companies.

In conclusion, Latin America's 5G network diffusion is a testament to the intricate balance of local and global forces shaping technological advancements. As countries strive to enhance connectivity and reap the benefits of the digital age, their strategies and outcomes will continue to be influenced by the evolving landscape of international relations, economic imperatives, and societal values. This dynamic environment will shape the region's telecommunications future, with lasting implications for regional development and global technological trends.

Notes

1 Tomás (2024a), online.
2 MINTIC (2023), online; Mayorga-Bohórquez et al. (2021).
3 Arias et al. (2021), p. 268.
4 Balbo and Cesarín (2020); Colombo et al. (2021); Quintana (2021).
5 Kaska et al. (2019); Bojić et al. (2021); Wang (2022).
6 *U.S. Envoy Returns to Brazil to Lobby against Huawei* (2020).
7 Arias et al. (2021), p. 262.
8 Government of Argentina (2021).
9 Interviews for this chapter were conducted between February 1, 2023, and May 10, 2023. The profiles of said interviewees range from civil society and academia to current government employees and former political figures. In total, eight interviews were conducted for the three initial review cases: Argentina, Chile and Colombia. Interviews are citated under APA guidelines.
10 Qualcomm (US) and Samsung (Korea) barely participate in the Latin American Market.
11 For example, Telefónica's profile at the SEC (US), mentions Ericsson, Google, and Huawei as technological partners (SEC, 2006).
12 See Appendix 1 for more details on the media outlets covered and number of tokens used as sample. Also, the exclusion of certain items and foreign media inserts of the analysis.
13 Alcalá Casillas (2019).
14 J. S., personal communication (June 3, 2024).
15 UNDP and World Bank (2021), pp. 5–9.

16 *World Internet Users Statistics and 2023 World Population Stats* (2023).
17 SUBTEL Chile (2022), online.
18 SUBTEL Chile (2023), online.
19 Arias et al. (2021), p. 261.
20 *World Internet Users Statistics and 2023 World Population Stats* (2023).
21 R. M. L., personal communication (March 16, 2023); T. P., personal communication (February 27, 2023).
22 OAS (2003), online.
23 Arias et al. (2021), p. 264.
24 ITU (n.d.), online.
25 Arias et al. (2021), p. 264.
26 Anonymous, personal communication (March 8, 2023).
27 Which include market distortion, the limitation of competition, the control over Chinese companies overseas, etc.
28 Strand Consult (2024), online.
29 Malena (2021), p. 4.
30 Stecklow and Dehghanpisheh (2020), online.
31 P. S., personal communication (January 18, 2023).
32 Hurwitz (2014).
33 T. P., personal communication (February 27, 2023).
34 Of sources related to technical areas within Telecommunications institutions across the region. (D. G. and A. O., personal communication, March 8, 2023; S. S. et al., personal communication, May 5, 2023).
35 T. P., personal communication (February 27, 2023).
36 Rozenwurcel and Lopez Fernández (2012), p. 8.
37 Carboni and Labate (2018).
38 Triviño et al. (2024).
39 *U.S. Envoy Returns to Brazil to Lobby against Huawei* (2020).
40 Li (2023), online.
41 D. G. and A. O., personal communication (March 8, 2023).
42 Ley 1341/ 2009 - Gestor Normativo - Función Pública (2009), article 2.
43 Cordero Pérez (2023), online.
44 Tomás (2024), online.
45 (M. P), personal communication, February 13, 2023; (P. S.), personal communication, January 18, 2023.
46 Some exceptions can be seen, like in the Chilean process, where an alternative Finish company, Borealnet, competed against the well-established providers.
47 *Claro* (2024).
48 *Countries and Emerging Business Units* (2021).
49 Farah and Richardson (2021), p. 9.
50 SUBTEL Chile (2020), p. 11.
51 M. P., personal communication (February 13, 2023); P. S., personal communication (January 18, 2023).
52 Flores (2019), online.
53 P. S., personal communication (January 18, 2023).
54 Cinco Días (2023), online.
55 The Corner (2023), online.
56 Editorial La República (2023), online.
57 Reuters (2023), online.
58 Marco (2024), online.
59 Such as Personal in Argentina or Entel Bolivia.
60 Arias et al. (2021), p. 262.
61 Government of Argentina (2021); Government of Ecuador (2024).

62 Pongratz (2024), online.
63 Strand Consult (2022), p. 55.
64 Strand Consult (2022), p. 50.
65 Strand Consult (2023), online.
66 This is based on over ten interviews to experts, public officials and politicians in Colombia, Argentina, Spain and Chile, conducted between January and July, 2023.
67 P. S., personal communication (January 18, 2023).
68 Editorial La República (2023), online.
69 T. P., personal communication (February 27, 2023).
70 J. S., personal communication (June 3, 2024).
71 J. S., personal communication (June 3, 2024).
72 Although independence can also be preferred, as in the case of Digital Television Standards.
73 M. P., personal communication (February 13, 2023).
74 M. P., personal communication (February 13, 2023).
75 Scheidt (2024), p. 20.
76 Strand Consult (2024), online.
77 Biblioteca del Congreso Nacional de Chile (2020), online.
78 Ellis (2022); M. P., personal communication (February 13, 2023).
79 M. L. R., personal communication (March 16, 2023).
80 MINTIC (2019), pp. 29–36.
81 MINTIC (2019), p. 37.
82 Please refer to the appendix 1 for details on volume and nature of these publications.
83 Putnam (1988), pp. 435–437.
84 Gourevitch (1978).
85 Ripsman (2009), p. 170.

Bibliography

5G: Sergio Massa hizo clinc caja y llegó la hora de la letra chica. (n.d.). LetraP. Retrieved May 21, 2024, from https://www.letrap.com.ar/economia/5g-sergio-massa-hizo-clinc-caja-y-llego-la-hora-la-letra-chica-n5404050

67 Markets Worldwide Have Commercial 5G Services. (n.d.). S&P Global. Retrieved May 21, 2024, from https://www.spglobal.com/marketintelligence/en/news-insights/research/67-markets-worldwide-have-commercial-5g-services

Aguiar, A. R. (2024, March 25). *Telefónica negocia con la SEC estadounidense una multa histórica y el motivo es haber mantenido relaciones con empresas chinas vetadas.* Business Insider España. https://www.businessinsider.es/eeuu-prepara-multa-historica-telefonica-negocios-chinos-1374752

Alcalá Casillas, M. G. (2019). Desigualdad en el acceso a internet en México y la afectación en el ejercicio del derecho humano a la información. *Nuevo Derecho, 15*(24), 55–70. https://doi.org/10.25057/2500672X.1122

Anonymous. (2023, March 8). *Interview with a Current Public Servant in Chile Who Requested Anonymity.* [In Person].

Arias, F., Salado, A., Medina, C., & Zambrano, M. (2021). 5G Technology Deployment in Latin America: An Analysis of Public Policy and Regulation Environment. *World Journal of Advanced Research and Reviews, 11*(3), 258–271. https://doi.org/10.30574/wjarr.2021.11.3.0457

Balbo, G., & Cesarín, S. M. (2020). *¿Guerra comercial, periferia tecnológica o tecnoimperialismo?: América Latina ante la competencia global en el sector de las telecomunicaciones.* ALADI. https://repositorio.aladi.org/handle/20.500.12909/30827

Bernal, L. S. (2024, March 13). *¿Cuáles son los avances de la red 5G en América Latina?* https://mobiletime.la/noticias/13/03/2024/avances-de-la-red-5g-en-america-latina/

Biblioteca del Congreso Nacional de Chile. (2020). *LEGISLATURA 368ª Sesión 78ª*. https://www.bcn.cl/laborparlamentaria/documento?id=692675

BNamericas—Industry Players Create Open RAN Lobby Group... (n.d.). BNamericas.Com. Retrieved June 7, 2024, from https://www.bnamericas.com/en/news/industry-players-create-open-ran-lobby-group-in-brazil

BNamericas—Radiografía de las redes 5G en Latinoamérica. (n.d.). BNamericas.com. Retrieved June 8, 2024, from https://www.bnamericas.com/es/noticias/radiografia-de-las-redes-5g-en-latinoamerica

Bojić, L. M., Đukanović, D., & Nikolić, N. G. (2021a). 5G as Geopolitical Power Struggle. *NBP – Nauka, Bezbednost, Policija, 26*(3), 25–47. https://www.ceeol.com/search/article-detail?id=1214759

Bojić, L. M., Đukanović, D., & Nikolić, N. G. (2021b, July 28). *Countries and Emerging Business Units*. Telefónica. https://www.telefonica.com/en/about-us/countries-emerging-business-units/

Cama-Pinto, D., Damas, M., Holgado-Terriza, J. A., Gómez-Mula, F., Calderin-Curtidor, A. C., Martínez-Lao, J., & Cama-Pinto, A. (2021). 5G Mobile Phone Network Introduction in Colombia. *Electronics, 10*(8), Article 8. https://doi.org/10.3390/electronics10080922

Cambero, F. (2021, March 22). *Chile Fast-Tracking 5G Roll-Out, but with Tight Rules on Security, Official Says*. Reuters. https://www.reuters.com/article/idUSKBN2BE11H/

Carboni, O. V., & Labate, C. (2018). América Latina por una red neutral: El principio de neutralidad en Chile y Brasil. *Revista FAMECOS, 25*(2), 28507. https://doi.org/10.15448/1980-3729.2018.2.28507

Cinco Días. (2023). *Telefónica y otras telecos dejan de lado a la china Huawei en el despliegue del 5G por los riesgos de seguridad nacional.* Empresas, Cinco Días. https://cincodias.elpais.com/companias/2023-07-10/telefonica-y-las-grandes-telecos-orillan-a-huawei-en-sus-nuevos-proyectos-tecnologicos.html

Claro. (2024). https://www.claro.com/

Colombo, S., López, M. P., & Vera, N. (2021). Tecnologías emergentes, poderes en competencia y regiones en disputa: América latina y el 5G en la contienda tecnológica entre China y Estados Unidos. *Estudos Internacionais: revista de relações internacionais da PUC Minas, 9*(1), Article 1. http://hdl.handle.net/11449/206412

Cordero Pérez, C. (2023, September 6). *Esto dice el decreto presidencial que prohibiría a China (Huawei) ser proveedor de 5G en el país.* El Financiero. https://www.elfinancierocr.com/ef-de-la-manana/gobierno-emitio-decreto-que-prohibe-compras-a/GPS76Z37CFALDBM7GPZFWEL57E/story/

Dano, M. (2018, September 10). *AT&T Names Samsung, Ericsson and Nokia as 5G Equipment Suppliers*. Fierce Network. https://www.fierce-network.com/5g/at-t-names-samsung-ericsson-and-nokia-as-5g-equipment-suppliers

Editorial La República S.A.S. (2023, December 7). *Telefónica prepara la salida de la china Huawei de las redes sus mercados más grandes*. Diario La República. https://www.larepublica.co/globoeconomia/telefonica-prepara-la-salida-de-la-china-huawei-de-las-redes-sus-mercados-mas-grandes-3763297

Ellis, E. (2022). *Colombia's Relationship with the PRC.* CSIS. https://www.csis.org/analysis/colombias-relationship-prc

Entel and Huawei, Jointly Complete the Commercial Deployment of Environment-Friendly TubeStar—Huawei Press Center. (2018). Huawei. https://www.huawei.com/en/news/2018/6/huawei-entel-commercial-tubestar

Farah, D., & Richardson, M. (2021). *The PRC's Changing Strategic Priorities in Latin America: From Soft Power to Sharp Power Competition* (Strategic Perspectives 37; pp. 1–43). Institute for National Strategic Studies INSS.

Figueroa, J. C. (2024). *JoseCardonaFigueroa/Sentiment-Analysis-Spanish* [R]. https://github.com/JoseCardonaFigueroa/sentiment-analysis-spanish (Original work published 2015).

Fletcher, B. (2019, April 16). *Battle for 5G RAN Market Share Neck and Neck in 2023: Strategy Analytics*. 5G Technology World. https://www.5gtechnologyworld.com/battle-for-5g-ran-market-share-neck-and-neck-in-2023-strategy-analytics/

Flores, C. (2019, December 17). *Telefónica quitará a Huawei de su 5G: Pasará a una estrategia de varios fabricantes de su red.* El Español. https://www.elespanol.com/omicrono/tecnologia/20191217/telefonica-quitara-huawei-pasara-estrategia-varios-fabricantes/452705569_0.html

G., D., & O., A. (2023, March 8). *Interview with Authorities at the Telecommunications Agency, Chile.* [In Person].

Gourevitch, P. (1978). The Second Image Reversed: The International Sources of Domestic Politics. *International Organization, 32*(4), 881–912. https://www.jstor.org/stable/2706180

Government of Argentina. (2021, July 23). *Becas Huawei—Centro G+T.* Argentina.gob.ar. https://www.argentina.gob.ar/jefatura/innovacion-publica/centro-gt/becas-huawei-centro-gt

Government of Ecuador. (2024, January 30). ARCOTEL Y HUAWEI CAPACITAN SOBRE TECNOLOGÍA 5G - Agencia de Regulación y Control de las Telecomunicaciones. *Agencia de Regulación y Control de las Telecomunicaciones - Promovemos el desarrollo armónico del sector de las telecomunicaciones, radio, televisión y las TIC, mediante la administración y regulación eficiente del espectro radioeléctrico y los servicios.* https://www.arcotel.gob.ec/arcotel-y-huawei-capacitan-sobre-tecnologia-5g/

Hardesty, L. (2024, January 22). *Exclusive: Samsung Gains on Nokia in Popularity with Operators, According to Recon Analytics.* https://www.fierce-network.com/wireless/exclusive-samsung-gains-nokia-popularity-operators-according-recon-analytics

Higuchi, K. (2016). A Two-Step Approach to Quantitative Content Analysis: KH Coder Tutorial Using Anne of Green Gables (Part I). *Ritsumeikan Social Sciences Review, 52*(3).

ITU. (n.d.). *About ITU.* ITU. Retrieved May 26, 2024, from https://www.itu.int:443/en/about/Pages/default.aspx

Jiménez, F. A. (2023, December 7). *Los chinos Huawei y ZTE ganarán más mercado con la subasta 5G en Colombia.* www.elcolombiano.com. https://www.elcolombiano.com/negocios/el-interes-chino-detras-de-la-subasta-5g-en-colombia-FK23293341

Kaska, K., Beckvard, H., & Minárik, T. (2019). Huawei, 5G and China as a Security Threat. *Nato Cooperative Cyber Defence Center of Excellence.*

Latinoamérica se mueve entre las sanciones de EU y los despliegues de 5G: Huawei. (n.d.). El Economista. Retrieved May 21, 2024, from https://www.eleconomista.com.mx/empresas/Latinoamerica-se-mueve-entre-las-sanciones-de-EU-y-los-despliegues-de-5G-Huawei-20230920-0032.html

Ley 1341 de 2009 - Gestor Normativo - Función Pública, 1341, Corte Constitucional (2009). https://www.funcionpublica.gov.co/eva/gestornormativo/norma.php?i=36913

Li, Y. (2023). *How Are Government's Decisions Affected by an External Power? The Case of Huawei 5G in Brazil.* https://hdl.handle.net/10438/33710

M. L., R. (2023, March 16). *Interview with a Former High-Level Politician, Colombia.* [Online].

Malena, J. (2021). The Extension of the Digital Silk Road to Latin America: Advantages and Potential Risks The DSR's Origins and Its Initial Development. *Council on Foreign Relations.*

Marco, A. (2024, March 25). *El supervisor bursátil de EEUU sanciona a Telefónica por su relación con China*. elconfidencial.com. https://www.elconfidencial.com/empresas/2024-03-25/cnmv-americana-impone-sancion-telefonica-relacion-china_3853562/

Mayorga-Bohórquez, M. Á., Estupiñán-Cuesta, E. P., & Martínez-Quintero, J. C. (2021). Diagnóstico de la situación actual de la tecnología 5G: Suramérica. *Visión Electrónica*, *15*(2), 272–283. https://doi.org/10.14483/22484728.17982

MINTIC. (2019). *Plan 5G Colombia El Futuro Digital es de Todos*. MINTIC, Government of Colombia

MINTIC. (2023). *Gobierno adjudica licencias de 4G - MINTIC - Vive Digital*. https://mintic.gov.co/portal/vivedigital/612/w3-article-2029.html

Mishra, Y. (2021, October 9). *Huawei Claims over 50 Percent of 5G Base Stations in China: Report*. Huawei Central. https://www.huaweicentral.com/huawei-claims-over-50-percent-of-5g-base-stations-in-china-report/

OAS. (2003). *Cybersecurity*. https://www.oas.org/ext/en/security/prog-cyber

OMC. (n.d.). *Contratación pública—Puerta de acceso*. Retrieved May 25, 2024, from https://www.wto.org/spanish/tratop_s/gproc_s/gproc_s.htm

P. M. (2023, February 13). *Interview with an Inter-American Development Bank Consultant* [In Person].

Pascal, R. (2023). *Market Landscape: RAN Vendors 2023*. Ericsson and Omdia Research

Pongratz, S. (2024, March 13). *Worldwide Telecom Equipment Market Slumps in 2023*. Dell'Oro Group. https://www.delloro.com/worldwide-telecom-equipment-market-slumps-in-2023/

Putnam, R. D. (1988). Diplomacy and Domestic Politics: The Logic of Two-Level Games. *International Organization*, *42*(3), 427–460. https://www.jstor.org/stable/2706785

Quintana, A. R. (2021). Latin American Countries Must Not Allow Huawei to Develop Their 5G Networks. *The Heritage Foundation*, *6041*.

Recon Analytics (x-China) Survey Reveals that Ericsson, Nokia and Samsung Are the Top RAN Vendors. (n.d.). Technology Blog. Retrieved May 21, 2024, from https://techblog.comsoc.org/2024/01/22/recon-analytics-x-china-survey-reveals-that-ericsson-nokia-and-samsung-are-the-top-ran-vendors/

Reuters. (2023, September 20). *Telefonica May Seek Damages in Event of Huawei Curbs in Germany*. Reuters. https://www.reuters.com/business/media-telecom/telefonica-may-seek-damages-event-huawei-curbs-germany-2023-09-20/

Ricart, R. J. (n.d.). *Geopolitical Aspects of the EU's Data Strategy*. Elcano Royal Institute. Retrieved May 23, 2024, from https://www.realinstitutoelcano.org/en/work-document/geopolitical-aspects-of-the-eus-data-strategy/

Ripsman, N. M. (2009). Neoclassical Realism and Domestic Interest Groups. In S. E. Lobell, N. M. Ripsman, & J. W. Taliaferro (Eds.), *Neoclassical Realism, the State, and Foreign Policy* (1st ed., pp. 170–193). Cambridge University Press. https://doi.org/10.1017/CBO9780511811869.006

Rozenwurcel, G., & Lopez Fernández, M. (2012). *Compras Públicas en América Latina y el Caribe ¿Internacionalizar o no internacionalizar? ¿Quién, cómo, cuándo?* Cátedra OMC FLACSO Argentina. https://flacso.org.ar/wp-content/uploads/2014/10/Compras-P%C3%BAblicas-en-Am%C3%A9rica-Latina-y-el-Caribe-09_FLA_OMC.pdf

S. J. (2024, June 3). *Telecommunications Consultant with Direct Connections with the International Market* [Online].

S. P. (2023, January 18). *Expert on China's Relations with Latin America* [Online].

S., S., R., H., & B., C. (2023, May 5). *Interview with the Colombian Communications Regulator's Commission*. [Online (Zoom)].

Scheidt, M. (2024). EU Policy on the Cybersecurity of 5G Networks. *Unit Cybersecurity Technology and Capacity Building DG CNECT, European Commission.*

SEC. (n.d.). *Telefónica Argentina.* Retrieved May 23, 2024, from https://www.sec.gov/Archives/edgar/data/932470/000110465911037168/a11-15564_120f.htm

SEC. (2006). *Telefónica, S.A.* https://www.sec.gov/Archives/edgar/data/814052/000095010306001082/dp02389_20f.htm

Stecklow, S., & Dehghanpisheh, B. (2020, July 14). *Exclusive: Huawei Hid Business Operation in Iran after Reuters Reported Links to CFO.* Reuters. https://www.reuters.com/article/idUSKBN23A19B/

Strand Consult. (2022). *The Market for 5G RAN in Europe: Share of Chinese and Non-Chinese Vendors in 31 European Countries* (pp. 1–64). Strand Consult. https://strandconsult.dk/the-market-for-5g-ran-in-europe-share-of-chinese-and-non-chinese-vendors-in-31-european-countries/

Strand Consult. (2023, June 15). *European Commission Presents Second 5G Toolbox Report – What It Means for EU Mobile Operators: Comparison of Rules to China.* Strand Consult. https://strandconsult.dk/european-commission-presents-second-5g-toolbox-report-what-it-means-for-eu-mobile-operators-comparison-of-rules-to-china/

Strand Consult. (2024, April 26). *There Are Many Myths about China and Its Technology Suppliers. The EU Published a Report You Should Read.* Strand Consult. https://strandconsult.dk/there-are-many-myths-about-china-and-its-technology-suppliers-the-eu-published-a-report-you-should-read/

Suárez, L. (2024, March 13). *¿Cuáles son los avances de la red 5G en América Latina?* https://mobiletime.la/noticias/13/03/2024/avances-de-la-red-5g-en-america-latina/

SUBTEL Chile. (2019). *Seminario Internacional sobre calidad de servicios de telecomunicaciones.* https://cdn.www.gob.pe/uploads/document/file/1231876/Subsecretaria_de_Telecomunicaciones_-_Visi%C3%B3n_y_perspectivas_sobre_QoS_y_QoE_en_el_contexto_de_5G_y_IoT.pdf

SUBTEL Chile. SUBTEL, Government of Chile (2020). *Concursos 5G, Norma técnica Ciberseguridad y Puerta Digital Asia Sudamérica.*

SUBTEL Chile. (2022). *Hogares con acceso a Internet fijo alcanzan el 67% y usuarios aumentan preferencia por redes de alta velocidad—Subsecretaría de Telecomunicaciones de Chile.* https://www.subtel.gob.cl/hogares-con-acceso-a-internet-fijo-alcanzan-el-67-y-usuarios-aumentan-preferencia-por-redes-de-alta-velocidad/

SUBTEL Chile. (2023). *Internet.* Subsecretaría de Telecomunicaciones de Chile. https://www.subtel.gob.cl/estudios-y-estadisticas/internet/

T. P. (2023, February 27). *Interview with a Former Economic Relations' Public Official, Chilean MOFA.* [Online].

The Corner. (2023). *Telefónica Prepares to Replace Huawei in Mobile Network, in Line with Western Veto of Chinese Companies.* The Corner. https://thecorner.eu/news-spain/telefonica-prepares-to-replace-huawei-in-mobile-network-in-line-with-western-veto-of-chinese-companies/112104/

Toda Latinoamérica es territorio chino gracias a Huawei. (n.d.). El Economista. Retrieved May 23, 2024, from https://www.eleconomista.com.mx/empresas/America-Latina-es-territorio-Huawei-20230805-0024.html

Tomás, J. P. (2024a, February 12). *Chile Modifies Calendar for Incoming 5G Spectrum Auction.* RCR Wireless News. https://www.rcrwireless.com/20240212/5g/chile-modifies-calendar-incoming-5g-spectrum-auction

Tomás, J. P. (2024b, February 14). *Costa Rican Court Suspends Exclusion of Huawei as a 5G Provider.* RCR Wireless News. https://www.rcrwireless.com/20240214/5g/costa-rican-court-suspends-exclusion-huawei-5g-provider

Triviño, R., Franco-Crespo, A., & Ochoa-Urrego, L. (2024). Network Neutrality Policies in the 5G South American Market. In G. F. Olmedo Cifuentes, D. G. Arcos Avilés, & H. V. Lara Padilla (Eds.), *Emerging Research in Intelligent Systems* (pp. 189–203). Springer Nature Switzerland. https://doi.org/10.1007/978-3-031-52255-0_14

Turcsányi, R., & Rühlig, T. (2023, August 4). *Evaluating Public Support for Chinese Vendors in Europe's 5G Infrastructure*. CEIAS. https://ceias.eu/evaluating-public-support-for-chinese-vendors-in-europes-5g-infrastructure/

UNDP & World Bank. (2021). Digitalisation and Productivity: In Search of the Holy Grail – Firm-Level Empirical Evidence from EU Countries. *OECD Economics Department Working Papers 1533; OECD Economics Department Working Papers* (Vol. 1533). https://doi.org/10.1787/5080f4b6-en

U.S. Envoy Returns to Brazil to Lobby against Huawei. (2020, November 9). CLBrief. https://www.clbrief.com/us-envoy-returns-do-brazil-to-lobby-against-huawei/

Wang, L. (2022). *China's Huawei in the Us-China Trade War in the Communications Sector Game* (pp. 485–497). https://doi.org/10.2991/aebmr.k.220603.078

What Is 5G? How Will It Transform Our World? (n.d.). Ericsson.Com. Retrieved May 23, 2024, from https://www.ericsson.com/en/5g

World Internet Users Statistics and 2023 World Population Stats. (2023). https://www.internetworldstats.com/stats.htm

10 How China Installed a Space Research Base in Argentina

Geopolitical Intrigue and Governance Challenges

Is the People's Liberation Army (PLA) operating a military base in south Argentina under the cover of a space exploration center? "Estación de Espacio Lejano" or *Deep Space* is a center for scientific research activities located in the Neuquén province, Argentina, since 2012—it has sparked doubts in recent years, as it was discovered that the Chinese Communist Party (CCP) army operated in the enclosure, turning it into a de facto military operation. As Beijing is becoming more straightforward in its military aims and intentions to gain military presence in the Asia-Pacific region and potentially in the Antarctic territories, some actors—including US officials—cast doubts over the objectives of maintaining a base with such characteristics in that location and the CCP's influence at a local and national level.[1]

These issues, some proven and some still under debate—are yet another example of the transfer of technology and scientific activities to the region. However, this is recognizably different from other cases analyzed in this book in which states discriminate among private companies that can be related to a country of origin, but mostly indirectly. In this case, the private sector is—primarily—not involved, and the outcomes are unarguably related to Argentinian security and the potential interests of China. Secondly, the assessment comes from a deal stroke between both governments, and the political situation in the country has widely varied since its opening in 2012, giving time for retrospective criticism of previous policies. The conditions under which the Neuquén territory was leased by President Cristina Fernandez (Kirschner) included the continuous use of 200 hectares for 50 years without economic compensation to the state or taxation—something linked to the agreement—among other political and logistical benefits granted by the Argentinian state. In addition, the Chinese staff entered the country under diplomatic schemes, making it difficult for national authorities to track their background or the nature of their work.[2]

In the vast arrangement of schemes through which countries transfer and receive technology, academic, public, and private scientific research is a core concern for states, as it involves transferring critical information that can be used for national security purposes or to extend a state's influence abroad. Therefore, universities and research centers are connected to these powers' soft power strategies and the process of knowledge acquisition. The case of this deep space station, one piece in the PRC's broader outer-space-related activities, is considered by many as a threat and a tool in the growing tensions between major powers in satellite and

DOI: 10.4324/9781003489450-12

telecommunications. Although both Europe as a bloc and the US conduct research related to space exploration in Latin America—and Washington also has a military presence—the main issue with the Neuquén base is the lack of transparency during its implementation process and operations.

In this chapter, I review concurring debates over the nature, controversy, and future of the Neuquén *Estación de Espacio Lejano*. I consult existing evidence and previous research on the matter and establish novel connections between the Neuquén case and the broader context of space exploration, threat perception, and military presence in the region. I also integrate the views of key actors from the Argentinian government, security experts, and sources relevant to the case, whose testimonies contribute to building a solid testimonial base to question the objectives of the PLA actions within the station. The results of this analysis are contrasted with other examples and reflected in this book's conclusions.

Latin America's Securitization Approach

Latin American states perceive international trends and threats differently from the rest of the world. The region has not been a historical hotspot for warfare generally outside the conflict zones during the First and Second World Wars. There are a few exceptions, such as the alleged connection between Mexico and Germany during the First World War that sparked the US' decision to enter said conflict and Cuba, as a satellite of the Union of Socialist Soviet Republics (USSR) during the Cold War—involved in the missile crisis in 1962. In recent years, countries in the region have taken different positions on confrontations abroad, with most siding with Ukraine in the case of Russia's aggression,[3] but with a divided position regarding the Palestine-Israel conflict and few concrete mentions when it comes to Asia Pacific. Regarding Taiwan and China's tensions, there are mixed visions,[4] and neutrality—or avoiding open support of one side—has generally prevailed. However, some countries, such as Chile, have begun publishing statements that contradict the PRC's position.[5]

Interstate war is also rare in the Latin American region and particularly in South America.[6] Although border disputes and discord exist between neighboring countries, and there are some examples of armed conflicts, these often do not last long and are seldom embedded in more significant ideological or international quarrels. In contrast, countries have been recognized for solving disputes in international courts or interacting with mediators,[7] including the Vatican. As a result, even though some countries have some of the highest murder rates in the world—higher than many warzones even—Latin America has remained relatively peaceful and safe and has hosted migrant waves escaping international conflicts for over a century.[8] The former, combined with the history of authoritarianism and secret police involvement in government affairs and persecution of dissidents across many states during the Cold War, has created a scenario in which securitization and intelligence efforts are also stigmatized.[9] (In)famously, operation Condor implied a shared security apparatus among anti-communist actors in Latin America,[10] and the

US has been known to have intervened in political processes making use of those same intelligence agencies.[11]

The end of the Cold War brought about economic liberalization and incentives for developing countries to open their markets to a set of nations willing to trade without too much concern over more significant international power competitions that did not directly affect national interests. Consequently, the region has increased efforts to maintain neutrality and avoid taking sides, which would technically benefit trade for these nations. However, this does not mean that states do not look at the region from a strategic perspective, either for its natural resources, geographical location, or its privileged access to the Antarctic territories.

The US has a presence through military bases[12] and agreements. By 2023, the only three countries below its southern border that did not have US military bases were Bolivia, Venezuela, and Mexico. Some of these locations are a result of historical events, such as the Guantanamo Bay area in Cuba. In contrast, others result from leasing or security cooperation agreements—something not surprising considering the penetration of the US presence abroad.[13]

Although the pursuit of neutrality has continued in the competition scenario between the US and China, this has become difficult as geopolitics and power balance emerge as priorities. The consensus is that countries will likely have to follow this international trend, which is detrimental to international liberalism and global economic aspirations, even more than protectionism itself.[14] As mentioned in previous chapters, Latin America is not exempted from changes in the balance of power, and a vital element of the competition between the US and China is technological dominance. In this context, terminology related to "hybrid warfare" and the geopolitical importance of telecommunications and satellites in this escalating rivalry is of uppermost importance, even though it hasn't been approached as such by Latin American states.[15]

Argentina has taken several steps toward a closer relationship with China, potentially due to its economic dependency and political conditions. President Néstor Kirschner arrived in power in 2003 and increased national efforts to become closer to Beijing. In 2014, Chinese executive leader Xi Jinping visited Argentina in a round of meetings that concluded with the Joint Declaration for the Establishment of a Strategic and Integral Association between China and Argentina (Asociación Estratégica Integral entre la República Argentina y la República Popular China), further consolidating the connections between both nations.[16] Argentina was excluded from access to credit schemes due to its inability to pay a US$100,000 billion bond and sought the help of Beijing. At the same time, the counterpart began negotiating the construction of the base.[17]

In contrast, President Javier Milei, who began to govern in 2023, is ideologically closer to the US but has also pointed to the need to maintain good relations with China. In this context, the review of the Neuquén space station documents and general operations is an outstanding example of the delicate balance and changes in threat perception for the nation. It also showcases how technologies can serve multiple purposes, with some raising more alarms than others for national and international political interests.

China's Space Aspirations and International Competition

The PRC launched its first human-crewed mission in 2003, surprising a world abandoning the idea of space exploration as a goal, which was quickly reflected in the cancelation of the space shuttle program in 2004.[18] The Chinese achievement was notable not only because few other countries had accomplished such a feat by that date but also because the country's economy was only beginning to gain a relevant international position. Yang Liwei became the first *taikonaut*, and Chinese society rapidly embraced becoming part of a new space race. Investing in space exploration, research, and development had several advantages for the CCP at the time. Military and strategic reasoning can be linked to the origins of the space program.[19] Still, success also gave them a better reputation internationally, helped create new industries that pushed economic metrics, and was an excellent way to boost nationalism across the population.[20] The PRC continues to progress toward becoming a leading competitor in the space industry, with its satellite system *Bei-Duo* being one of its most important projects.[21] It also plans to build a space station, reach the dark side of the moon, and launch missions beyond the lunar orbit soon.[22] Beijing also wants to increase its private space industry and significantly lower costs in rocket technology, mainly to gain flexibility and shorten the time and effort of satellite replacements.[23]

Space rapidly became an essential element of the competition between the US and China, even though it wasn't on the radar a decade ago.[24] Both countries have recognized the need to prepare space for an interstate conflict, as telecommunications largely depend on satellites, and this can be a target to be disrupted. As a result, the development of space-related weaponry and military activities connected to space research has become part of the technological debate.[25]

During the Cold War, the USSR competed directly with the US in achieving milestones in exploration. Today, space competition and the industry related to its capacity are not exclusive to great powers, and the reasons for gaining the capacity to access orbit are political and economic. The satellite industry has become widely profitable—with the space economy expected to reach one trillion dollars in 2040[26]—and because the cost of manufacturing and launching space operations has been exponentially reduced since man reached the moon. These conditions allowed private actors to compete directly in the market[27] and for regional powers such as India or Japan to establish their strategic objectives in space. For example, Japan is looking at potential ways to mine asteroids—with successful samples taken in 2019,[28] and India has launched rockets with impressive results.[29]

States, international organizations, and the private sector have strong motivations to control space-related activities. Communication and interconnectivity, two critical pillars of a globalized world, are increasingly dependent on the interaction of satellites in low Earth orbit and radio frequencies operating on the planet. Governments may be interested in preferential access to space or critical locations for space exploration and other strategic purposes. For some experts and government agents, this could be why China was interested in building a space research facility in the arid Argentinian Patagonia.[30]

The Neuquén Base: Background and Controversy

Countries in the Southern Hemisphere are also interested in space, although this is only sometimes accompanied by the capacity to achieve significant milestones. In this context, states have used their geographic position and climate to collaborate with space research endeavors. In Chile, astronomic research outputs are over the world's average. The country has an advantageous position due to its dry and clear desert in the north, and this can be linked with the installation of observatories and astronomy-related facilities.[31] Argentina also has a privileged position for space observation in some country locations, such as Mendoza Province, where the European Union opened the ESA Deep Space 3 (DSA3) "to support European interplanetary exploration space missions."[32] Other Argentinian localities have been considered for various projects,[33] sometimes competing with options in Chile.

Argentina has long dwelled with financial constraints, making China an alternative to access loans and other funding mechanisms. In this context, a deal was negotiated—under certain degrees of secrecy—between the Argentinian and China during Cristina Fernandez de Kirschner's government[34] to establish the Neuquén *Estación de Espacio Lejano*. The station was conceived to support space missions in the more comprehensive Chinese Lunar Exploration Program (CLEP) and to communicate with satellites and spacecraft—in their efforts to study the dark side of the moon and continue their Mars missions.[35] To that end, a 35-m diameter parabolic antenna was installed at the base as part of an investment estimated at US$300 million, but that could have been around 1 billion US dollars.[36]

The deep space station was established in the vicinity of *Bajada del Agrio,* a relatively secluded town of less than a thousand people. As Figure 10.1 shows, the facilities are completely isolated to the public. The desertic region provided access to a dry environment and clear sky—not so contaminated with artificial light—and where operations could be conducted without much intervention or contact with the rest of the country. For some locals, representatives, and elected officials in the province of Neuquén, this was seen as giving away the state's sovereignty without proper consultation.[37] Many others were worried about the military presence in the base due to the close links with the Chinese army and the increasing international political tensions. However, despite concerns, the station began operations in 2017[38] and was completely operational by 2018.

Estación de Espacio Lejano was developed following a series of agreements by several organizations from both states, including the China Satellite Launch & Tracking Control General (CLTC), the Argentinian National Commission for Space Activities (CONAE), and the Neuquén provincial government. In June 2012, the Minister of Development for the Neuquén Province,[39] Elso Leandro Bedoya, signed a decree granting the CONAE the right to lend a plot of land of 200 hectares for 50 years—with the possibility of an extension—to the Chinese counterpart without compensation or the payment of rent. The text indicates that the land corresponds to "Paraje Quintuco Pilmathue, Loncupué Department, Neuquén," and that it is meant for data acquisition and control of China's National Space Administration

Figure 10.1 This is a picture of the deep space research station in Neuquén, showing the facilities in the middle of the dry Patagonian Desert. The picture was kindly provided by IBI Consultants and was taken as part of their field work activities on the field of regional security, 2024.

(CNSA) institution for their Moon and Mars missions.[40] A month later, in July 2012, CONAE signed the final deal with the Chinese CLTC and the Neuquén province, specifying the coordinates of the project, agreeing on 15 articles, and establishing the conditions under which the station was allowed to operate.[41]

A new document was signed in April 2014. The Argentinian government and the PRC agreed that all activities and trade related to the establishment, construction, and operation of the deep space station by the Chinese counterpart would be exempt from taxation and tariffs and that Argentina could not interfere or supervise the operation. This is controversial because it removes potential benefits for

Argentina, which allows Beijing to utilize its territories without clear benefits or control mechanisms.[42]

From the beginning, the base raised concerns due to the possible undisclosed uses the PRC could give that strategic location.[43] Security experts and government agents agree that these alleged activities include the disruption of telecommunications, military surveillance, espionage, and cyber defense activities, which would be against the terms under which the base was conceived and agreed to be built.[44] One reason for this ongoing concern is that the Chinese PLA directly supervises the CLTC, which is part of what Beijing considers security strategical institutions. Another vital element is the alleged secrecy and confidentiality behind the agreement.[45] Overall, there seems to be little direct benefit for the Argentinian government regarding revenue, research capabilities, and the development of the isolated Neuquén area, so the general belief is that Beijing made secondary or indirect promises in exchange for the deep space base.[46]

Concerns persisted over a potential dual use of the base in the following years despite these agreements' apparent compliance with Argentinian law. In an increasingly tense international scenario, civil society, politicians, figures from the Argentinian military, and foreign diplomats continued to pressure the government to take concrete action to ensure that the PLA would not violate the conditions proposed in the open agreement.[47] In 2016, the Mauricio Macri government (2015–2019), who opposed the Kirschner or Peronist administration that signed the deal, commended his Chancellor, Susana Malcorra, to reevaluate the conditions of the agreements. Consequently, in September 2016, the Argentinian representative traveled to Hangzhou, China, to meet her counterpart Wang Yi. Both authorities released a joint agreement reassuring the public that the space station for observation in Neuquén was exclusively for *peaceful* activities.[48] As a result, the "Space Agreement" between Argentina and China allows military personnel to be present at the station as long as they are strictly part of *peaceful* scientific research. However, there aren't any mechanisms to corroborate the nature of the PLA operations within the perimeter granted to the Chinese entity. In the 2014 agreement (article 3.2), the Argentinian government agreed not to interfere in the space station, making it unlikely for local authorities to verify if the activities complied with the terms.[49] Moreover, an annex to the agreement between the two nations requires Argentina to secure the radioelectric spectrum between 10 and 50 GHz in a radium a hundred kilometers surrounding the station and guaranteeing that there wouldn't be any remote supervision of the base.[50] Regarding security considerations, the concerns are the capacity to gather peripherical telecommunications data from the station and the triangulation of Neuquén with other Chinese facilities.[51] Figure 10.2 provides a closer look of the infrastructure, with two shots of the antennas.

As a result, there is ongoing criticism of this 2016 agreement revision since it needed to establish precise control mechanisms or pressure China to honor the original contract. Even though CONAE, ENACOM, and the Neuquén province visit the premises and can access certain documents, there is doubt over how enforceable the agreement is. Moreover, the state is already evaluating potential

Figure 10.2 This is a picture of the deep space research station in Neuquén, showing a closer shot of the antennas. The picture was kindly provided by IBI Consultants and was taken as part of their field work activities on the field of regional security, 2024.

consequences for Argentina as tensions between the US and China grow in the region.[52]

Argentina has since grown closer to the PRC, which can also be correlated with a deteriorated economic scenario and the state's diplomatic objectives. In 2018, under Macri's administration, both governments signed a "Joint Action Plan" in which both parties expressed their intentions to deepen bilateral cooperation regarding the Neuquén station.[53]

Looking at these agreements, testimonies from sources, and the broader scope of evidence, the conclusion is that it is generally agreed that the process of installing the Neuquén station and the consequent actions by both states do not violate domestic regulations. However, the presence of specific clauses and articles that allow for the non-disclosure of activities in the area where the PLA is situated, the station's potential to interfere in telecommunications, and the lack of transparency reported by locals, alongside escalating tensions between the US and China, have made this a sensitive case that has garnered significant criticism. Unlike other cases in this book, the implemented technology is not the main reason for contingency but the perceptions of threat and interests from the involved parties.

The Neuquén Base as Part of the Wider Power Competition

The US has been vocal about the risks involved in this project and considers it a hindrance in future bilateral relations between Buenos Aires and Washington. Again, the concerns were mostly related to Argentina's lack of knowledge about PLA activities in the area and the objectives behind this operation. The US has increased its interests in the country, opening to sponsor several infrastructure and technical projects, including a multi-purpose port in Ushuaia that could eventually be used for defense.[54] In 2019, the former commander of US Southern Command, US Navy Admiral Craig Faller, visited Argentina to meet local defense authorities.[55] In a report before the Senate Armed Services Committee on the Neuquén issue earlier that year, Faller mentioned that:

> Beijing could be in violation of the terms of the agreement with Argentina to only conduct civilian activities and may have the ability to monitor and potentially target US, allied, and partner space activities.[56]

Although the issue of the Neuquén base remained an underlying discussion, the public debate resurfaced in 2024, shortly after Javier Milei began his administration. In March of that year, American Ambassador Marc Stanley declared he was surprised that the Argentinian government allowed PLA personnel to be present at the "deep space" station[57] without supervision. Soon after, General Laura Richardson, the head of the US Southern Command (SouthCom), arrived in Argentina in April 2024 to emphasize her country's position on the growing presence of China in Latin America. She also raised concerns about the Neuquén deep space station.[58]

That same month, after continuous criticism from security experts on the factual background and operations of the Chinese station, the Argentinian government decided to put all contracts, agreements, and public and secret documents related to the Neuquén deep space Station under review.[59] President Milei has openly stated his intention for Argentina to develop a closer relationship with the US, and taking action on the Neuquén base can be interpreted as a step toward improving this connection.[60]

However, there are reasons to criticize the speculation over the base, pointing out the lack of concrete evidence of intervention and the US' objectives over the issue, as well as to assess the existence of threats of the Neuquén states not only from potential risks but also by the existing evidence of collaboration and benefits to Argentinian institutions and research. Those who support the latter point out that this influence might affect local perceptions and threat perception, as well as hinder Argentinian interests and opportunities for collaboration in science and technology since CONAE has access to space exploration resources as stipulated in the agreement, and other Argentinian universities and institutions have taken advantage of this condition.[61] Needless to say, China has openly denied any military involvement and directly condemns the US alleged campaign to spread fear over peaceful research activities.[62]

Consequently, Argentina is a direct point of geopolitical confrontation between both countries, which is reflected in their actions and interests. The often-mentioned goal of maintaining neutrality or non-alignment in a global escalation of tensions is challenging for states heavily intertwined with the US and China. In other words, the presence of Chinese military personnel in Argentina has begun to pressure the country to make political decisions beyond those directly related to scientific and astronomical research. Moreover, even if Argentina has the power to put an end to the agreement—it has to do it five years in advance—the administration is aware that direct action against the PRC will have consequences that most governments would prefer to avoid, predominantly economical, as Beijing could decide to retaliate by cutting trade or stop investment.[63]

Other Space Research Activities in South America

Neuquén is not the only example of the region's Chinese space and astronomical endeavors. In Chile, an international hub for celestial and outer-space research, PRC-related institutions have also shown interest and conducted research for at least a decade. The South American Astronomy Center, CASSACA, is a joint project between Chilean and Chinese institutions.[64] The center has operated since 2013, and in November 2023, both countries celebrated the 6th annual bilateral conference for Astronomy.[65] According to Chinese state media, the center is instrumental in coordinating Chinese visiting researchers to use telescopes located in Chile and constructing telescopes in Chile with Chinese funding.[66] China also announced significant investments in building an observatory in *Cerro Ventarrones*. In 2023, the Chilean Catholic Northern University (UCN), the Chinese National Astronomical Observatory (NAOC), and the Chinese Academy of Science (CAS) announced the development of the *Ventarrones* observatory. They declared that the project would last at least ten years.[67]

There isn't a direct link with the Chinese military in Chinese space-related projects within Chile, as in the Neuquén case, and they are widely accepted as a first and foremost academic effort. However, for some South American security consultants and experts, this project is an example of a wider strategy and the PRC's aspirations to become a space leader, either in missions or the development of technology.[68]

Conclusions and Further Questions

Establishing the "Estación de Espacio Lejano" in Neuquén, Argentina, is controversial due to the direct link between the PLA and the institutions involved in developing the project. By situating a critical component of its space program in Argentina, China has extended its scientific capabilities and potentially created a platform for intelligence gathering and military operations. Considering that the CLTC is under the authority of the PLA, it's reasonable to assume that the data it collects could come under the purview of China's defense apparatus and potentially serve military objectives. However, this doesn't necessarily mean that the data will be used for aggressive or combat-related purposes since China is not the only country that conducts research under the supervision of military personnel.[69]

In this case, the transfer of space technology involves the increasing relationship between China and Argentina. After reviewing documents and talking to relevant sources, it becomes clear that the Neuquén station did not provide the near region with development or economic benefits due to the tax and tariff exemption granted to the PRC counterpart. Although it does provide some benefits for the local research community, the focus of the controversy lies at the intersection between geopolitics and national sovereignty. Since the CCP has emphasized its space program while the US has raised concerns over PRC presence in the region, Argentina struggles to take a position on the topic, which is also affected by consecutive changes in government that draw closer to either one of the two powers. In other words, the perception of the Neuquén base as a threat to national sovereignty seems to be part of a more significant geopolitical issue, which public officials acknowledge.[70]

The dual-use nature of space technology underscores the blurred lines between peaceful scientific endeavors and strategic military advantages. This duality challenges host nations and the international community to regulate and monitor such installations to ensure compliance with international norms and national interests. It is also relevant due to the non-interference and exclusive use of spectrum terms stipulated in the documents signed before the completion of the deep space station.

From Argentina's perspective, the agreement to host the Chinese facility highlights significant governance challenges. The lack of transparency and minimal oversight over the Neuquén station operations result from agreed terms that are compliant with Argentinian regulations. However, it is generally agreed that the initial agreement struck under the administration of former President Cristina Fernández de Kirchner lacked sufficient safeguards to ensure Argentine sovereignty and oversight.[71] Subsequent administrations have struggled to impose stricter controls or renegotiate terms, reaching revisions that attempt to ensure the peaceful use of the installation, but that—according to some experts—is not enough considering China's track record in other parts of the world, including the South China Sea.[72]

The US apprehension about the militarization of space and the expansion of Chinese influence in its hemisphere has led to increased scrutiny and diplomatic tensions. Countries in South America need to be aware of their geopolitical position and how to navigate these two states' expressed and underlying interests. Argentina is also aware of the conflict and power dynamics. As the Buenos Aires government faces an economic crisis, they need to secure stable relations to avoid repercussions that could affect their development and domestic political scenario.[73]

The Argentine experience with the Neuquén station is also a cautionary tale for other nations engaging in similar agreements with significant powers. It highlights the need for comprehensive legal and regulatory frameworks that can adapt to evolving geopolitical landscapes. Nations must ensure that agreements are transparent, include stringent oversight mechanisms, and allow regular audits and inspections to safeguard national interests. Additionally, engaging in multilateral dialogues and partnerships can provide a counterbalance to bilateral agreements that may be skewed in favor of more powerful nations.

In conclusion, the "Estación de Espacio Lejano" case illustrates the complex interplay of science, military strategy, and international relations. It underscores the importance of vigilance and adaptability in international agreements, particularly those involving sensitive technologies and strategic infrastructure. As space becomes an increasingly contested domain, the experiences of nations like Argentina will provide valuable lessons in navigating the challenges of global governance, sovereignty, and security in the 21st century.

Moving forward, Argentina and other nations must reassess their strategic priorities and strengthen their institutional capacities to manage such high-stakes collaborations effectively. This includes fostering greater transparency, enhancing domestic regulatory frameworks, and engaging in proactive diplomacy to mitigate potential risks. The evolving dynamics of global power require a nuanced approach that balances national interests with the broader imperatives of international peace and security. By learning from the Neuquén experience, nations can better prepare for the complexities of future technological and geopolitical challenges.

Notes

1 P. Z., personal communication (June 14, 2024).
2 "*Qué hay y qué más puede haber en la inquietante base de China en Neuquén*" (2024).
3 Sanahuja et al. (2022).
4 Considering that several countries in Central America, and Paraguay, still recognize the ROC over the PRC as the diplomatic representative of China.
5 Minrel (2024).
6 Martín (2006).
7 Franchi et al. (2017).
8 Durand and Massey (2010); Nobbs-Thiessen (2021), pp. 949–953.
9 This takeaway comes from an anonymous interview with a former security agent, and I cannot disclose the name, location, date, or any further details of this meeting.
10 As described by McSherry (1999).
11 Kornbluh (2007).
12 Lindsay-Poland (2005), online.
13 Cemeri (2023), online.
14 Rodrik (2023), online; Hurrell (2024).
15 J. S., personal communication (June 3, 2024).
16 Castillo-Argañarás (2022).
17 Londoño (2018), online.
18 Hammer (2012).
19 *China's Vision for Space*. (n.d.).
20 Seedhouse (2010), pp. 3–9.

21 Government of China (2021), online.
22 Goswami (2023), online.
23 Goswami (2023).
24 Nacht (2013), online.
25 Goswami (2023), online; P. Z., personal communication (June 14, 2024).
26 *A New Space Economy on the Edge of Liftoff* (n.d.); *What Is the Space Economy?* (2019).
27 Weinzierl and Sarang (2021).
28 *Hayabusa2 and Ryugu* (n.d.).
29 *The History and Motivations behind India's Growing Space Program* (2024).
30 P. Z., personal communication (June 14, 2024); R. F., personal communication (June 16, 2024).
31 Cortes et al. (2018).
32 Durst et al. (2019), p. 118.
33 Aubé et al. (2014).
34 Londoño (2018).
35 P. Z., personal communication (June 14, 2024).
36 R. F., personal communication (June 16, 2024).
37 Londoño (2018); González Manriquez (n.d.).
38 Grigioni (2024), online.
39 Considering that Argentina is a Federal State, the chain of events and regulation between local and national affairs varies depending on the nature, size and objectives of a project.
40 Resolución 391/2012 (2012).
41 *Acuerdo de Cooperación En El Marco Del Programa Chino de Exploración de La Luna Entre El China Satellite Launch and Tracking Control General (CLTC) y La Comisión Nacional de Actividades Espaciales (CONAE) de La República Argentina Para Establecer Instalaciones de Seguimiento Terrestre, Comando y Adquisición de Datos, Incluida Una Antena Para Investigación Del Espacio Lejano, En La Provincia de Neuquén, Argentina* (2012).
42 Castillo-Argañarás (2022), pp. 119–122.
43 Padinger (2024, April 6); Pérez (2016, March 17).
44 P. Z., personal communication (June 14, 2024); R. F., personal communication (June 16, 2024).
45 Moreno Crosgrove (n.d.).
46 P. Z., personal communication (June 14, 2024).
47 R. F., personal communication (June 16, 2024).
48 Government of Argentina (2018).
49 Castillo-Argañarás (2022), pp. 119–122.
50 Montes de Oca and Monge Molina (2019).
51 P. Z., personal communication (June 14, 2024).
52 R. F., personal communication (June 16, 2024).
53 Castillo-Argañarás (2022).
54 Chaves (2024), online.
55 U.S. Southern Command (2019).
56 Faller (2019), p. 7.
57 Rosemberg (2024), online.
58 Mathus Ruiz (2024).
59 Chaves (2024), online.
60 Buenos Aires Times (2024a); Buenos Aires Times (2024b).
61 R. F., personal communication (June 16, 2024).
62 Global Times (2024).
63 R. F., personal communication (June 16, 2024).
64 *CASSACA – CAS Centro Sudamericano para la Astronomía* (2024)
65 Upon reviewing the website in Spanish, English and Chinese, I found a series of inconsistencies that might not be relevant for this chapter, but that might show lack

of transparency. For example, the Spanish version shows little information about the objectives of the center, while the staff and researchers in charge in the Chinese and the English version are different.

66 CGTN (2023).

67 *NAOC de China y la UCN dieron el vamos a proyecto astronómico conjunto en Cerro Ventarrones « Noticias UCN al día – Universidad Católica del Norte (*2023).

68 P. Z., personal communication (June 14, 2024).

69 Caro (2024).

70 R. F., personal communication (June 16, 2024).

71 Castillo-Argañarás (2022), pp. 119–122.

72 P. Z., personal communication (June 14, 2024).

73 R. F., personal communication (June 16, 2024).

Bibliography

A New Space Economy on the Edge of Liftoff. (n.d.). Morgan Stanley. Retrieved June 17, 2024, from https://www.morganstanley.com/Themes/global-space-economy

Acuerdo de Cooperación En El Marco Del Programa Chino de Exploración de La Luna Entre El China Satellite Launch and Tracking Control General (CLTC) y La Comisión Nacional de Actividades Espaciales (CONAE) de La República Argentina Para Establecer Instalaciones de Seguimiento Terrestre, Comando y Adquisición de Datos, Incluida Una Antena Para Investigación Del Espacio Lejano, En La Provincia de Neuquén, Argentina, 10715, 10715 (2012).

Aubé, M., Fortin, N., Turcotte, S., García, B., Mancilla, A., & Maya, J. (2014). Evaluation of the Sky Brightness at Two Argentinian Astronomical Sites. *Publications of the Astronomical Society of the Pacific*, *126*(945), 1068–1077. https://doi.org/10.1086/679227

Buenos Aires Times. (2024a, April 2). *Government Considers Inspection of Chinese Space Station in Neuquén.* Buenos Aires Times. https://www.batimes.com.ar/news/argentina/government-assesses-whether-to-inspect-chinese-base-in-neuquen.phtml

Buenos Aires Times. (2024b, April 5). *Milei: 'We Must Strengthen Strategic Alliance with United States.'* Buenos Aires Times. https://www.batimes.com.ar/news/argentina/milei-we-must-strengthen-strategic-alliance-with-united-states.phtml

Caro, C. J. V. (2024). *The Patagonian Enigma: China's Deep Space Station in Argentina.* The Diplomat. https://thediplomat.com/2024/01/the-patagonian-enigma-chinas-deep-space-station-in-argentina/

CASSACA – CAS Centro Sudamericano para la Astronomía. (2024, February 29). http://www.cassaca.org/en/

Castillo-Argañarás, L.-F. (2022). La estación espacial china en la Patagonia: Una aproximación desde el derecho internacional. *URVIO. Revista Latinoamericana de Estudios de Seguridad*, *33*, 109–124. https://doi.org/10.17141/urvio.33.2022.5298

Cemeri. (2023). *What Are the Latin American Countries with the Most US Military Bases?* Cemeri.Org. https://cemeri.org/en/mapas/m-bases-militares-eeuu-americalatina-cu

CGTN. (2023). *Chile cuenta con el primer Centro Conjunto China-Chile de Astronomía.* https://espanol.cgtn.com/news/2024-04-30/1785197392186089474/index.html

Chaves, P. F. (2024, April 9). *El Gobierno revisará los contratos de la estación china en Neuquén pero por ahora no habrá visita ni inspección.* infobae. https://www.infobae.com/politica/2024/04/09/el-gobierno-revisara-los-contratos-de-la-estacion-china-en-neuquen-pero-por-ahora-no-habra-visita-ni-inspeccion/

China's Vision for Space. (n.d.). The National Bureau of Asian Research (NBR). Retrieved June 17, 2024, from https://www.nbr.org/publication/chinas-vision-for-space/

Cortes, R., Depoortere, D., & Malaver, L. (2018). Astronomy in Chile: Assessment of Scientific Productivity through a Bibliometric Analysis. *EPJ Web of Conferences, 186*, 05002. https://doi.org/10.1051/epjconf/201818605002

Durand, J., & Massey, D. S. (2010). New World Orders: Continuities and Changes in Latin American Migration. *The Annals of the American Academy of Political and Social Science*, *630*(1), 20–52. https://doi.org/10.1177/0002716210368102

Durst, S., Safonova, M., Paolantonio, S., Colazo, M. E., & Li, G. (2019). Galaxy Forum South America-Argentina 2020. *Proceedings of the International Astronomical Union*, *15*(S367), 116–119. https://doi.org/10.1017/S1743921321000831

F., R. (2024, June 16). *Interview with a Public Official Serving Argentina's Internal Affairs on Defense Matters.* [Online].

Faller, Adm. C. S. (2019). *Posture Statement of Admiral Craig S. Faller, Commander, United States Southern Command before the 116th Congress* [Unclassified]. Senate Armed Services Committee. https://es.slideshare.net/LeonardoAttuch2/southcom-2019-posturestatementfinal#1

Franchi, T., Migon, E. X. F. G., & Villarreal, R. X. J. (2017). Taxonomy of Interstate Conflicts: Is South America a Peaceful Region? *Brazilian Political Science Review*, *11*, e0008. https://doi.org/10.1590/1981-3821201700020008

Global Times. (n.d.). *Fallacies Spread by US Southern Command Commander on China-Argentina Deep Space Exploration Cooperation "Absurd, Lack Basic Respect": Embassy.* Global Times. Retrieved June 19, 2024, from https://www.globaltimes.cn/page/202404/1310525.shtml

González Manriquez, L. E. (n.d.). *La estación «Espacio Lejano» de Neuquén: Pieza clave en la guerra de satélites del firmamento austral.* CADAL.ORG. Retrieved June 11, 2024, from https://www.cadal.org/publicaciones/informes/?id=15576

Goswami, N. (2023). *China's Space Program in 2023: Taking Stock.* The Diplomat. https://thediplomat.com/2023/12/chinas-space-program-in-2023-taking-stock/

Government of Argentina. (2016). *Argentina y China reafirman el uso pacífico de la Estación Espacial de Neuquén.* https://www.cancilleria.gob.ar/es/actualidad/comunicados/argentina-y-china-reafirman-el-uso-pacifico-de-la-estacion-espacial-de

Government of Argentina. (2018, May 30). *Estación CLTC - CONAE-NEUQUEN.* Argentina.gob.ar. https://www.argentina.gob.ar/ciencia/conae/centros-y-estaciones/estacion-cltc-conae-neuquen

Government of China. (2021). *Full Text: China's Space Program: A 2021 Perspective.* ENGLISH.GOV.CN. https://english.www.gov.cn/archive/whitepaper/202201/28/content_WS61f35b3dc6d09c94e48a467a.html

Grigioni, J. P. (2024, April 5). *Estación Espacial China en Neuquén: Transparencia y seguridad jurídica.* Neuquén Informa. https://www.neuqueninforma.gob.ar/estacion-espacial-china-en-neuquen-transparencia-y-seguridad-juridica/

Hammer, M. (2012). The World without the Space Shuttle. *More Important than Money: Sleep.* The Science in Society Review, Arizona State University (vol. 47). 47–50., Inc.

Hayabusa2 and Ryugu. (n.d.). Astromaterials Science Research Group | ISAS. Retrieved June 17, 2024, from https://curation.isas.jaxa.jp/

Hurrell, A. (2024). Geopolitics and Global Economic Governance. *Oxford Review of Economic Policy*, *40*(2), 220–233. https://doi.org/10.1093/oxrep/grae013

Kornbluh, P. (2007). *Declassifying U.S. Intervention in Chile.* NACLA. https://nacla.org/article/declassifying-us-intervention-chile

Lindsay-Poland, J. (2005, October 5). *U.S. Military Bases in Latin America and the Caribbean.* Institute for Policy Studies. https://fpif.org/us_military_bases_in_latin_america_and_the_caribbean/

Londoño, E. (2018, July 28). *Desde una estación espacial en Argentina, China expande su presencia en Latinoamérica*. The New York Times. https://www.nytimes.com/es/2018/07/28/espanol/america-latina/china-america-latina-argentina.html

Martín, F. E. (2006). *Militarist Peace in South America*. Palgrave Macmillan US. https://doi.org/10.1057/9781403983589

Mathus Ruiz, R. (2024, April 3). *Laura Richardson, la general de EE.UU. que trabaja día y noche para frenar el avance de China en América Latina*. LA NACION. https://www.lanacion.com.ar/politica/laura-richardson-la-general-de-eeuu-que-trabaja-dia-y-noche-para-frenar-el-avance-de-china-en-nid03042024/

McSherry, J. P. (1999). Operation Condor: Clandestine Inter-American System. *Social Justice*, *26*(4), 144. https://www.public.asu.edu/~idcmt/terror.pdf

Minrel. (2024). *Comunicado por elecciones en Taiwán*. Minrel. https://minrel.gob.cl/noticias-anteriores/comunicado-por-elecciones-en-taiwan

Montes de Oca, I., & Monge Molina, J. (2019). *Argentina quedó atrapada en medio de una disputa militar por el espacio entre los Estados Unidos y China*. Infobae. https://www.infobae.com/politica/2019/02/15/argentina-quedo-atrapada-en-medio-de-una-disputa-militar-por-el-espacio-entre-los-estados-unidos-y-china/

Moreno Crosgrove, N. (n.d.). *Global Affairs and Strategic Studies: Facultad de Derecho*. Global Affairs and Strategic Studies. Retrieved June 8, 2024, from https://www.unav.edu/web/global-affairs/detalle/-/blogs/la-controvertida-estacion-espacial-de-china-en-la-patagonia

Nacht, M. (2013). *The United States and China in Space: Cooperation, Competition, or Both?* (Anti-Satellite Weapons, Deterrence and Sino-American Space Relations, pp. 101–112). Stimson Center. https://www.jstor.org/stable/resrep10894.10

NAOC de China y la UCN dieron el vamos a proyecto astronómico conjunto en Cerro Ventarrones. (n.d.). Noticias UCN al día – Universidad Católica del Norte. Retrieved June 20, 2024, from https://www.noticias.ucn.cl/noticias/estudiantes/naoc-de-china-y-la-ucn-dieron-el-vamos-a-proyecto-astronomico-conjunto-en-cerro-ventarrones/

Nobbs-Thiessen, B. (2021). New Waves of Immigration and Departure in Modern Latin America. *Latin American Research Review*, *56*(4), 946–957. https://doi.org/10.25222/larr.1669

Padinger, G. (2024, April 6). *¿Qué sabemos de la estación del Espacio Lejano que China opera en la Patagonia argentina?* CNN. https://cnnespanol.cnn.com/2024/04/06/que-sabemos-estacion-espacio-lejano-china-patagonia-argentina-orix/

Pérez, C. (2016, March 17). *Lo que se sabe de la misteriosa base que China está construyendo en la Patagonia argentina*. BBC News Mundo. https://www.bbc.com/mundo/noticias/2016/03/160317_misteriosa_base_china_patagonia_argentina_lb

Qué hay y qué más puede haber en la inquietante base de China en Neuquén. (2024, April 10). Memo. https://www.memo.com.ar/poder/base-china-neuquen/

Resolución 391/2012, N° 4340-001302/12, Neuquén Province (2012).

Rodrik, D. (2023, September 6). *The Global Economy's Real Enemy Is Geopolitics, Not Protectionism*. Project Syndicate. https://www.project-syndicate.org/commentary/global-economy-biggest-risk-is-geopolitics-not-protectionism-by-dani-rodrik-2023-09

Rosemberg, J. (2024, April 1). *Marc Stanley: "Me sorprende que la Argentina permita que Fuerzas Armadas chinas operen en Neuquén."* LA NACION. https://www.lanacion.com.ar/politica/marc-stanley-nadie-puede-decir-que-la-amenaza-del-terrorismo-se-haya-ido-de-la-argentina-nid31032024/

S., J. (2024, June 3). *Telecommunications Consultant with Direct Connections with the International Market* [Online].

Sanahuja, J. A., Stefanoni, P., & Verdes Montenegro, F. J. (2022). *América Latina frente al 24-F ucraniano: Entre la tradición diplomática y las tensiones políticas*. Fundación Carolina. https://doi.org/10.33960/issn-e.1885-9119.DT62

Seedhouse, E. (2010). *The New Space Race: China vs. USA*. Springer Science & Business Media.

The History and Motivations Behind India's Growing Space Program. (2024). The Planetary Society. https://www.planetary.org/articles/history-motivations-indias-space-program

U.S. Southern Command. (2019). *Adm. Faller Visits Argentina*. https://www.southcom.mil/MEDIA/IMAGERY/igphoto/2002151179/

Weinzierl, M., & Sarang, M. (2021). *The Commercial Space Age Is Here*. Harvard Business Review. https://hbr.org/2021/02/the-commercial-space-age-is-here

What Is the Space Economy? (2019). https://space-economy.esa.int/article/33/what-is-the-space-economy#_ftnref1

Z., P. (2024, June 14). *Interview with a Security Consultant Focused on Latin America* [Online].

11 Conclusions

A Final Review of the Role of Geopolitics, Economic Incentives, Values, and Reputation in Technology Transfers in Latin America

Technologies are ever-changing and have increasingly become crucial to societal interactions.[1] Reliance on digitalization, networks, and new frontier research helps societies improve quality of life, increase access to services, and secure broader equality in access to information. As ethically challenging as some technologies—such as generative AI or autonomous machines—are, progress and development can seldom be defined without considering scientific and technological research outputs.[2] This also affects international relations and research related to that field, which I explored across the chapters of this work.

States have numerous incentives to create, disseminate, acquire, and compete in technological fields beyond the military sphere.[3] Power relations between states have long been influenced by unequal material capacities,[4] including innovation in that field. However, current discussions over technology transfers also include other critical areas—and infrastructure—that often involve public and private operations, such as the telecommunications sector.[5]

In this book, I introduced a critical yet widely unexplored area of research in international relations studies: the effects of technology transfers on international relations and the geopolitical landscape of Latin America. This endeavor considered an examination of the various dimensions of international relations theory, technology dissemination, and the intricate interrelationship between developing countries, middle powers, and global superpowers, with a particular focus on China and the US—the focus of much concern in a region shadowed by a history of colonialism, lack of technology development, extractive economies and a convoluted political history.[6]

Understanding that this region includes a varied and unequal assortment of countries, I proposed a series of questions regarding the outcomes of decision-making processes, particularly in cases where the PRC is often portrayed as a competitor to the US in narrative terms or by decision-makers.[7] In some cases, these public discussions based on a geopolitical discourse minimized the presence of other competitors and some regional leaders. In others, an overdue security-related debate on the effects of state-led technology in the region remains unexplored or is not seen as urgent by regional actors.

I established various objectives and frameworks that elaborate on existing international relations theory related to technology, the role of states, and the conditions

DOI: 10.4324/9781003489450-13

under which states seek to disseminate innovation. I also considered historical and philosophical reasoning to provide a cross-disciplinary framework to explain the counterintuitive outcomes of some technology transfer processes in Latin America. This was incorporated in each chapter as required by the context and the results of the presented case studies.

There were three feasible lines of reasoning for the government to evaluate providers, sometimes in direct negotiations and others in public tenders. The first is economic incentives or decisions taken only because of financial reasons, the second is shared concerns about geopolitics—particularly the idea of a rivalry between the US and the PRC and a potential new Cold War scenario—and the third is value-related limitations, such as the region's tradition on data privacy, or reputational factors associated with shared ethical and political frameworks. I acknowledge that decision-makers' reasons can overlap in some of the analyzed cases. Still, I found that the process in which a provider has discarded or selected is primarily straightforward or at least agreed upon by researchers and experts.[8]

In this conclusion chapter, I summarize the contributions of this book and analyze the results by comparing and contrasting the outputs with the presented theory. I also briefly elaborate on the concepts central to this proposal, including the idea of a so-called "Cold War" competition in technology across the Latin American region and the role of the US and China. Finally, I present possible new lines of research based on these findings and limitations that might be covered in future works.

Theoretical Contributions and Implications

One overarching objective of this book is to provide a nuanced understanding of how technology and power dynamics shape the international order, especially in the context of Latin America. This implies the incorporation of cross-disciplinary tools and analysis of existing theories to frame the results of each chapter and the technology that is reviewed within them. Generally, I found that international relations theories offer distinct perspectives on the dissemination of technology and the power dynamics involved, providing valuable insights but also facing limitations in fully capturing the complexity of modern international relations.

Realism emphasizes the anarchic nature of the international system and the perpetual power struggle among states.[9] It highlights the role of technology in enhancing military capabilities and national security. However, realism often overlooks the influence of non-state actors and the importance of domestic politics in shaping international decisions.[10] The theory's focus on state-centric analysis can be restrictive when addressing the multifaceted nature of technology transfers and their implications for international relations. *Constructivism* offers a more nuanced approach by considering the role of social norms, identities, and relationships in shaping state behavior. It emphasizes the importance of understanding the social context in which international interactions occur.[11] Nevertheless, constructivism may fall short in addressing the empirical uncertainties and material constraints[12] that influence state decisions, particularly in the realm of technology transfers.

Liberalism focuses on the potential for cooperation and interdependence among states, driven by shared economic and political interests. It highlights the role of international institutions and norms in facilitating collaboration.[13] Nonetheless, liberalism can be overly optimistic, often underestimating the impact of power politics and states' strategic use of technology to advance their national interests.

In recognizing the strengths and limitations of these traditional theories, I propose *neoclassical realism* as a more comprehensive framework to encompass the nuances of this problem. Neoclassical realism integrates systemic and domestic factors—public opinion, economic conditions, and political alliances—acknowledging the influence of international trends and internal political dynamics on state behavior. This approach is particularly relevant for understanding the complex interactions of middle powers, which must navigate between global superpowers while addressing their domestic conditions. By recalling Robert Putman's "two-level games,"[14] this framework offers a more comprehensive understanding of the interplay between global trends and domestic factors. It guides the analysis beyond state-centric perspectives and considers the broader social, economic, and political contexts that shape international interactions.

Moreover, this argument's strength is based on the general theoretical framework and the application of cross-disciplinary knowledge and other theories. In that sense, I also resorted to Joseph Nye's notions of power[15] to understand the connection between private companies and the embedded reputation of countries. Technology has become a significant component of soft power as countries seek to enhance their global image and influence through technological prowess. Producing and exporting cutting-edge technology with a broad societal impact can bolster a country's reputation and attract international partnerships. This dynamic is evident in the competition between China and the US, where technological leadership is intertwined with broader geopolitical ambitions.

I also consulted the theory of backwardness advantage to explain the apparent position of China and the literature on harmonization and the diffusion of technology, which frames the context and decision-making processes explored in the cases of this book.[16] Overall, disseminating technology, significantly advanced technologies like 5G networks, play a crucial role in contemporary international relations. Understandably, adopting new technologies is driven by economic and political incentives.[17] States may choose technology providers based on cost, quality, and potential financial benefits. However, political considerations often complicate these decisions, such as the desire to strengthen alliances or avoid dependency on a single superpower. The interplay between economic incentives and political dynamics is a recurring theme in the case studies presented in this book. In other words, technology enhances economic development and connectivity and serves as a tool for geopolitical influence and soft power projection. This is done by the association between a country's reputation, policies, and geopolitical objectives and the companies that are assessed, discriminated and selected at a state-level for projects in Latin American territories.

States are critically concerned about the security implications of technology transfers. Advanced technologies, such as 5G networks, carry the potential for surveillance, espionage, and cyber threats. Therefore, the choice of technology providers can reflect broader security alliances and strategic considerations.

A Proposal on the Nature of the Interacting Actors in Technology Transfers across Latin America

Much of the discourse on technology transfer in Latin America centers on the rivalry between the US and China and the implications of this conflict for the region.[18] This competition concerns economic interests, ideological influence, and strategic dominance. This, combined with the emergence of shared behavior between like-minded countries, has prompted the discussion of whether this process can be considered a new form of Cold War, something I explored through the cases and elaborated upon later in this conclusion chapter.

I found that both countries have engaged in extensive efforts to expand their technological footprint in the region, particularly China, leveraging its technological advancements and public diplomacy strategies to gain influence. This includes establishing research centers through direct negotiation with local governments, providing scholarships or training programs, and engaging in infrastructure projects aimed at embedding Chinese technology and standards within Latin American markets. The US, on the other hand, has been cautious of China's growing influence, particularly in sensitive sectors like telecommunications and cybersecurity—while using its diplomatic missions to disseminate these concerns across local actors, decision-makers, and the media.[19] This geopolitical tension reflects a broader global trend where technology increasingly becomes a battleground for influence and control. Adopting technologies such as 5G networks, facial recognition systems, and cybersecurity solutions often involves not just a choice of technology but also an alignment with the values and geopolitical stance of the provider nations. In this context, the concept of technological sovereignty—where nations strive to control their technological infrastructure and data—has gained prominence as countries become more aware of the geopolitical implications of technology. This is an angle that will be worthy of further research and discussion.

In addition to those concerns, the issue of soft power, where cultural influence and reputation play crucial roles, is increasingly pertinent to this debate. Its cultural and educational initiatives show China's use of technology as a soft power tool. At the same time, the US leverages its technological dominance to promote its values and standards globally.

However, I found that although this is an essential point of analysis to explain the outputs of decision-making processes related to the dissemination of technology in Latin America, the exclusive focus on the US-China technology rivalry inevitably shadows the participation of other actors that have also successfully exercised influence and transferred technology to the region. In reality, countries such as Japan, South Korea, France, and Israel play critical roles in the technological landscape of Latin America—as regional actors, like Chile and Argentina, or even more notably Brazil, act as referents for other governments that often follow their decision-making and technical adjudication processes. These countries offer alternatives that are sometimes preferred due to concerns over security, governance, or alignment with international standards. For example, European companies have been significant players in the deployment of identification, data management, and data-processing technologies, often favored for their alignment with the European Union's stringent data protection regulations. Actually, the General Data Protection Regulation (GDPR) of the European Union serves as a benchmark for many Latin American countries, which are increasingly aligning their laws with these standards.[20] This trend highlights the influence of regulatory frameworks in shaping technological adoption and underscores the importance of harmonizing standards to facilitate smoother technology transfer and integration. It also reinforces the idea of a direct influence of values and reputation in decision-making in Latin America since GDPR is seen as a benchmark for regional regulation of data.

Some of these countries are considered middle powers. Defined by their behavior and strategic choices rather than sheer resources, middle powers play a crucial role in the global dissemination of technology.[21] These states must navigate a complex landscape of international relations, balancing their interactions with superpowers while pursuing their national interests. Middle powers often act as mediators, coalition builders, and norm entrepreneurs, leveraging their diplomatic skills and strategic positions. In this book, I considered the approaches of countries such as Japan, Sweden, and Brazil in their interactions with global superpowers. Again, neoclassical realism is a valuable framework for understanding the behavior of middle powers in the context of technology transfers. I proposed that by integrating systemic and domestic variables, this approach captures the complexity of their decision-making processes. This is seen in examples in which local issues affect a country's position on adopting and transferring technology, which is particularly strong in sensitive innovation and using citizens' data by foreign companies.

The Intersection of Technology, Innovation, and Global Dynamics in Latin America

As part of the global south, composed of emerging economies, considerable internet penetration rates, traditionally a region in which the US exercises historical influence, and with growing Chinese influence—Latin America is likely the most relevant region to study when comparing decision-making, power dynamics, and international norms related to technology transfers. The results of this book serve as a referent to understand future processes concerning both the adoption of new technologies or other political processes in the same region but can also be used to analyze cases in other countries that face comparable challenges related to innovation and international relations.

I found that the interplay between technology and global dynamics, particularly in Latin America, continuously changes as innovation poses new governance challenges. The analysis of this region's technological trajectory involves the implementation of standards intertwined with international norms and the prevalence of regional entities such as the Organization of American States (OAS)—which influences the region's approach to new technologies, cybersecurity, and the concept of privacy.[22] In addition, Latin America exhibits characteristics unique to its political reality and tradition. The idea of *Habeas Data* used to frame users' rights over their private information[23] has gained attention from policymakers and civil society to describe Latin America's approach to the issue.

The adoption of technologies depends widely on domestic and international political factors. In some cases, such as Venezuela's implementation of an identity card, direct negotiations and international political alliances with other illiberal countries were central to the outcome. In other cases, countries established public tender schemes to decide which provider would be selected for nationwide technological infrastructure projects. In these cases, the price was not the only consideration in the offers.

This book's chapters are divided according to different technologies that impact human development, societal dynamics, security-related affairs, and the economic behavior of communities. In each case study, I also explore each case's different implications on the general argument. For example, when analyzing the diffusion of digital television in South America, the role of Japan and Brazil, in contrast to the European and American alternatives, portrays the importance of harmonization for specific countries in the region. Moreover, in this case, as well as in facial recognition and the 5G spectrum adjudication process, Brazil's decision-making outputs are often considered by other countries in the neighborhood.[24] An international relations expert based in Brazil, whom I consulted on this matter, explained that this outcome is to be expected due to the size of the Brazilian economy. Additionally, Brazil has well-prepared human capital that can compete at a lower operating cost.[25]

One key shared characteristic across the case studies is the strategic use of technology transfer as a tool for broader geopolitical influence. China and the US view technology as an economic asset and a means to extend their influence and secure their interests in Latin America. However, the approaches taken by these two powers differ significantly, leading to varied outcomes depending on the local context and the specific technologies involved. Latin American decision-makers and public officials have been making efforts to steer clear of being involved in the power struggle between China and the US, as they mentioned in interviews for this book, by advocating for neutrality. In democratic countries in the region, there is a tendency to recognize the rivalry in private conversations but refrain from addressing it in public statements to avoid escalating tensions with either Beijing or Washington.

As a result, there aren't strict screening mechanisms to exclude any company from participating in public tenders. Exceptions of this include Costa Rica's recent decision to limit 5G deployment so companies from China could not participate, and that became highly controversial[26] for its political implications. Nevertheless, other cases have proven the need to consider and incorporate strategic thinking into decision-making. The lack of coordination between national security, foreign affairs, citizens' identification, and data-processing public departments in Chile resulted in the Chilean state's cancellation of a public tender for the management system of passports and ID cards.[27] This decision was made due to security concerns regarding a Chinese company, Aisino, being responsible for managing citizens' data. Additionally, there were concerns about the potential geopolitical implications for the relationship between Washington and Santiago.

Despite the intention of neutrality and the sources' technical assessment of these processes, the cases explored in this book demonstrate that countries in the region are not immune to the escalating tensions between the US and China. Additionally, private companies from middle powers have a significant presence. This presence can be attributed to regional and international leadership, soft power, and economic competitiveness. In some cases, certain countries have a long history of technology transfer with the region, making their systems and standards the logical option to reduce costs. This is due to the harmonization of existing digital and physical infrastructure and the reputation they have built over the years.

Finally, it is necessary to address the institutional weaknesses that might directly affect the outcome of a decision-making process. Direct negotiations between states—something that China tries repeatedly and that the US is not a stranger to, as proven by the presented cases—can lead to a lack of competition and an inability to compare possible providers. In addition, corruption is a tangible issue in Latin America, and it can directly affect the outcome of a process, making it impossible to analyze under the same conditions as clear, transparent, and open processes. Although I incorporate some of these cases in this book, further research should attempt to distinguish the level of democratization, transparency, and nature of the transfer of a specific technology. In this book, I found it necessary to show how these direct negotiations sometimes benefit a specific power, even when their reputation is compromised.

The Economic Reasoning behind Technology Transfers

When deciding which provider to deploy a particular strategic technology that requires international negotiations, some democratic states in Latin America rely on highly technical assessments and non-discrimination to establish the frameworks under which potential offerors will participate. For several of these processes, the combination of the economic offer—the price—and the results of the technical qualities of a proposal can determine the outcome. The results generally do not always consider the cheapest or most economically "beneficial" option. This is because offers that are too low can be questioned or even dismissed when they are not realistic or might be distorted to gain an advantage.[28]

In other cases—such as spectrum auctions—the agencies that present the project prefer to rely on "beauty contests" rather than public tenders.[29] This system, popularized by regional leaders and widely used in Japan,[30] not only addresses the price of an offer but allows for broader creativity and tailored solutions that might include a state's specific needs.[31] For example, when extending a new feature in telecommunications, governments might be interested in said technology reaching remote areas because of the economic impact it might bring in the future. In other words, these contests focus on the "public benefit" of the expected project, but this also means that the expected economic sustainability of a proposal might be affected.

According to experts and public officials, in an ideal scenario, decisions should not be based on the country of origin but on the actual proposal. However, assessments sometimes fail to consider possible consequences that are not directly technical or economic. Two critical issues are the geopolitical context and miscommunication across government agencies. In short, it has become increasingly difficult to answer to both powers' interests in the region despite this being feasible and generally accepted in the early 2000s.

In most of the reviewed cases, the results of the decision-making processes did not only depend on fair and solely economic competition between companies but on associated factors, which included geopolitical considerations, political decisions, and prioritization of values. When negotiating directly with local governments, companies—sometimes supported by their country's origins' diplomatic missions—have a considerable advantage if the local environment or institutional weaknesses create the conditions for it. For example, Chinese companies have a clear advantage when promoting technology transfer to Venezuela, an established ally in Latin America. In Argentina, the lack of access to international credit and the country's changing political landscape might be related to the conditions under which the Neuquén deep space base came to fruition. In other cases, companies like the Brazilian-American Johnson consortium, Japanese NEC, or Finland's Nokia might negotiate with local governments to test certain technologies directly, although this is seldom controversial.

In short, after reviewing all cases, Latin American states implement certain technologies transferred from other countries not only because of the "price" of the competing offers, even if constrained economic conditions exist. Harmonization can lead to specific mirroring of policies and decisions,[32] but this is not the rule. Experts agree there is a lack of regional coordination[33] and high inequality in technology adoption. This is evidenced by factors such as the differences in internet penetration and the disparities in the stages of adopting new controversial technology, such as the often-discussed 5G network deployment in Latin America.

Geopolitical Rivalry and Security Concerns

The cases I explore in this book show that public officials and experts are aware of how the increasing competition between the US and PRC impacts their domestic policies and decision-making. Generally, countries will not publicize their preference for one provider and will try to keep application requirements at a minimum. This is part of their institutional culture and is referred to as "neutrality" or "neutralidad" in Spanish.[34] Technical entities in democratic countries attempt to reduce the influence of this rivalry in the process of transferring technology by maintaining said neutrality and non-discrimination principles—while political actors and experts communicate an increasing concern over the need to consider the security implications of acquiring innovation from companies that would eventually have access to susceptible data and the privacy of citizens.

In addition, Latin American media outlets tend to expose the ongoing rivalry between the US and China regarding the transfer of telecommunications technology, somewhat mirroring and shaping public opinion on the matter. Although this is not surprising, I found that Huawei is the most-mentioned company when discussing security geopolitics over the 5G deployment process.[35] Other companies that are also involved in 5G deployment, such as Nokia, Ericsson, or Qualcomm, are comparatively absent—if not ignored—in the public debate. This could be the result of a successful soft-power strategy on the Chinese side to publicize its competitor, but considering the adjacent words used to refer to Huawei, it is more likely that the securitization narrative of Western and liberal countries, such as Australia, Japan, or the US, has transpired to regional media.

In this context, the testimonies from experts and officials involved in the issue show that public opinion does have an impact and can even pressure states to revisit the results of specific processes or to discuss security issues openly. In the case of the Chilean identification and passport system awarded to the Chinese company Aisino, the process was retracted due to several factors, including procedural issues with the original application and internal discussions between Chilean entities, particularly concerning foreign affairs and internal affairs. The phenomenon was slightly different in the case of the transpacific cable, which was meant to connect the Chinese coast with Chile and become a door for internet transit to South America. Chilean decision-makers saw in Huawei's conditioning of participating companies and overall pressure an uninvited restriction to the transparency of the process. Also, the capacity of the Japanese alternative to integrate the offer with Australia,[36] and consequently the Google offer to take over the project and build the submarine cable, became a more attractive alternative that, I argue, is also inherently related to geopolitical factors.

This is not always the case. In Ecuador and Argentina, the media has amplified concerns over the participation of PRC companies in surveillance and space technology, respectively, but the state's reaction has been more cautious. China is still a hugely important economic partner for many regional nations, and democratic governments prefer to avoid unnecessary conflicts. In contrast, when a country applies limitations, there is also wide criticism, as in the Costa Rican case.[37] In less democratic states or governments that share certain ideological features with the CCP, geopolitical alliances are more prominent. China has helped Venezuela build an identification system to control its population in the shape of the Carnet de la Patria, and a Cuban company provided Argentina with surveillance systems.

Some may argue that this cautiousness and lack of screening mechanisms have prompted an expansion of CCP political influence in the region. A security expert I interviewed during my research was emphatic about regional organizations and leading states' role in this process, especially the China-CELAC summit and the Belt and Road initiative and their links with academic institutions and local governments that are easier to negotiate with directly.[38]

China's strategy in Latin America focuses on offering comprehensive technology solutions integrated with its broader Belt and Road Initiative (BRI). China positions itself as a critical partner for Latin American development by providing technology and the financing and infrastructure needed to implement it. This approach is evident in the widespread adoption of Chinese telecommunications technology and the growing presence of Chinese tech companies in the region. However, this strategy also raises concerns about digital sovereignty and the potential for technological dependence. The reliance on Chinese technology could lead to a situation where Latin American countries find themselves constrained by the technical standards and governance models imposed by China, limiting their ability to operate independently in the digital sphere.

The US approach has been deemed reactive rather than proactive, focusing on countering Chinese influence rather than offering a compelling vision for technology-driven development in Latin America. In a considerable share of the examples in this book, American companies are not even participating in adjudication processes or negotiating the transfer of alternative technologies to China. Even in those cases, the US has explicitly attempted to warn—and even pressure—Latin American governments, as evidenced by Washington's behavior over the Neuquén Espacio Lejano base in Argentina or the case of the Chilean passport public tender. Some may also argue that this reactive stance limits the US's ability to fully engage with Latin American countries looking for more immediate and tangible benefits from their technology partnerships.

Values, Privacy, Data Integrity, and International Norms

As mentioned before in this work, there are three main intervening variables that affect possible decision-making outputs in the realm of technology transfers in the region. The first two are economic and security factors. However, the cases also reveal how value-related considerations—such as data privacy, human rights, and political alignments—influence technology transfer decisions. These considerations often create ethical dilemmas for Latin American states, particularly when engaging with technologies with dual-use potential or implications for civil liberties.

Shared norms and standards within the region and the international landscape also appear as a compelling argument to establish the frameworks under which states will invite private companies to deploy technologies. The International Telecommunications Union (ITU) is relevant in the context of internet diffusion in the region, but many more organizations currently shape the landscape. In the OAS, both the cybersecurity framework[39] and the general resolutions achieved in the realm of technology are considered by region when establishing a normative baseline for the operation of specific innovations that deal with sensitive data or have a considerable social and economic impact. Again, the Costa Rican decision to limit the participation of Chinese companies in its 5G process is an example of how specific Latin American countries could apply international norms to shape the landscape of their technology deployment.

Countries have also updated and adapted laws and regulatory frameworks in concordance with the higher traffic of data and new technologies that considerably impact governance and society. Again, I need to repeat that the vital referent for this was the European General Data Protection Regulation (GDPR), which some countries in Latin America attempted to mirror in their legal systems, either because of the value of user-level rights or the possibility of harmonizing systems to gain economic benefits from that. Interestingly, Argentina has had a more developed legal privacy system that was homologated with the GDPR at its release. It serves as a clear example of a tradition of data-protection values that can be found in other countries in the region. When only looking at the deployment of 5G networks in Latin America, public documents reveal that, for the cases analyzed, states used both regional leaders and Western countries' implementation processes as a benchmark to establish their own spectrum auction frameworks.[40] However, despite the efforts to update data and privacy regulations, these issues, as well as citizens' cybersecurity and human rights considerations, are not often mentioned by the media or the official documents when referring to technologies that need to be incorporated from foreign firms. This could be because these issues might be considered secondary to the technical needs of technology deployment, directly irrelevant, or because countries prefer to maintain non-discriminatory principles over debating or even screening issues that are not economic or purely technical. This book provides an introductory review of this issue, but further research can expand on the benefits and possible downsides of screening companies because of values such as data privacy or specific human rights considerations—while the media could also provide more in-depth coverage of these issues.

Finally, in my discussions with politicians and off-the-record sources, I found that some bureaucrats explicitly pointed to concerns about the reputation of illiberal countries, which was—to a certain extent—reflected in the media discourse. Middle powers enjoy a less controversial image, making companies based in locations such as Japan or Europe somewhat attractive in comparison. This can be attributed to their capacity to build up soft power, cultural influence, and a better image, which takes advantage of the rivalry between the US and China.

In that context, some Latin American countries' reluctance to fully embrace Chinese technology can be partly attributed to concerns over the potential erosion of democratic norms[41] and human rights concerns associated with Chinese surveillance technologies. However, as shown by the cases presented in this book,[42] this is secondary to technical and geopolitical factors. Moreover, countries with increasing democratic challenges or that are directly authoritarian do not seem to consider these shared values in their decisions.

The New "Cold War" in Technology Transfers to Latin America

Rather than only considering the US and China's rivalry in the outcomes of technology-related policies in Latin America, this book's critical point of debate is to frame the discussion under the concept of a "new Cold War." Throughout the chapters of this book, I analyze some cases in which neither the US nor the PRC successfully lobbied or pushed for their companies to be chosen in open processes. I also reviewed the discussions around possible geopolitical conditions under which policymakers and public officials assessed a specific technology. Overall, there seem to be ideological reasons for alliances and decisions taken at a state level in the region. In countries like Venezuela, political alliances with China and isolation—to an extent—shape the technology incorporated in the country.[43] In countries in which the ruling government's ideology aligns with either of the powers at a point in time, some decisions can be connected to those particular conditions. For example, Jair Bolsonaro's government explicitly criticized having a Chinese company providing 5G networks in Brazil.[44] The Argentinian Kirschner government was closer to the left when negotiating to implement the Neuquén deep space station, whereas the Milei government has expressed concerns over those operations.[45]

This pattern can be interpreted in favor of the idea of alliances in the context of a new Cold War, in which these two powers, the US and China, have divided the world in compliance with values, geopolitical interests, and economic pressure. If this is the case, then even if American companies don't compete in a process, Washington's influence can still be present when a company from a state that shares the values of the democratic West is selected or negotiates its entry into a Latin American market through a public contract. The US has consistently framed its technology transfer policies around protecting against Chinese influence in the region. The extent to which these security concerns are prioritized varies, leading to divergent outcomes across different states.

Beijing's allies and political intentions are more evident, and the CCP is investing resources in enlarging its influence, even in the form of regional organizations and the consolidation of the Belt and Road Initiative—although this is more related to infrastructure. China's approach emphasizes economic cooperation and mutual development, often downplaying the security implications of its technology transfers. For instance, Ecuador's engagement with Chinese companies for its national surveillance system highlights how China packages its technology transfers as development assistance, appealing to governments looking to modernize their infrastructure without stringent security conditions attached.[46]

However, the reasons for these outcomes are not always consistent, which makes it troublesome to declare that there is an established ideological alliance—in the case of the diffusion of digital television in the region, Europe and the US pushed for their systems to be deployed across the region. Still, Brazilian influence successfully cooperated with Japan and transferred the Japanese system to most of South America.[47] In the case of the Neuquén deep space base in Argentina, the US directly warned the Argentinian government; institutions have deemed that the agreement with Beijing is within the law despite specific confidentiality clauses. In Chile, several cases show that the country is crucial for both the US and China's projection to South America. The CCP's indirect pressure to negotiate exclusivity when building the transpacific cable caused pushback on the Chilean counterpart,[48] which can be the result of concerns over sovereignty as well as it's stronger ideological affinity with a democratic West. Moreover, the conditions for this new Cold War are very different from the factors that characterized the escalation of tensions between the US and the Soviet Union after the Second World War. For example—but not limited to—incompatible economic systems, direct participation in conflicts, and the threat of nuclear proliferation.

The arguments favoring classifying these dynamics within a new Cold War framework recognize that the ideological debate is not between Chinese and American values but more of a democratic, like-minded world and cooperation of illiberal and often authoritarian countries that attempt to change international norms. This can be recognized in technology transfers in the region, as China has a close relationship with local authoritarian nations, which materializes in a more comprehensive international scheme. Although Russia doesn't have a significant technological presence in the region—except, perhaps from the dissemination of the Sputnik vaccine during the COVID-19 pandemic—the Kremlin does cooperate with Beijing in the transformation of values that embed the international system, attempting to push for sovereignty over freedom, neutrality on the internet, and citizens' rights in the digital world.[49] On the side of nations that defend the International Liberal Order, technologies and regulations seem to be directed at reducing the counterpart's—but in mostly Chinese—influence due to security concerns and maintaining the fundamental values of the existing international norms. Several of the cases analyzed in this book follow this logic. In the case of the transpacific cable between Chile and Asia, it is clear that the Japanese proposal could have benefited from the shared values between the interacting nations, including Australia—while the Chinese insistence caused concerns in some parts of the Chilean government. In the case of the passport public tender concerning the same country, Washington's direct influence and the perceived benefit of good diplomatic relations with the US also played into the decision to review the outcome of the contest, which later revealed flaws and even alleged corruption in the Chinese proposal. In the 5G spectrum auctions, several politicians expressed the need to acknowledge the broader geopolitical context that caused controversy over Huawei's offers in the region.

Therefore, although the idea of a new Cold War between the US and China might be attractive because it would simplify and explain the rising rivalry between those two countries, I find it necessary to deepen the theoretical context in which this concept is being utilized to explain phenomena in international relations to prevent unrealistic extrapolations or interpretations. This debate will continue and might evolve as geopolitical alliances become influential and international norms are shaped.

Future Research and Limitations

In this book, I have analyzed cases of technology transfers to Latin American countries in various schemes, ranging from local direct negotiations to open public tenders and auctions that affect more than one country at once. The types of technologies analyzed also had different levels of securitization or implications regarding social consequences and impacts on international relations' dynamics. Without a doubt, the diffusion of innovation has a profound impact on citizens in the region. It thus requires a comprehensive analysis of the interplay between international relations theory, technology dissemination, and the behavior of middle powers. Faster connectivity, access to digital services, overall security, and scientific development are only a few ways modern technologies are attractive tools to boost development. When discussing technology transfer, it's crucial to consider aspects such as normative frameworks and debates over sovereignty, autonomy, security, and ethics. Upon reviewing various cases, I discovered that international norms governing innovation implementation influence the domestic processes of South American countries. This, in turn, impacts how private actors involved in technology supply are evaluated. As a result, companies' involvement in these processes is not purely driven by economic incentives but also by a desire to cultivate public trust and, in the case of the country of origin, to demonstrate soft power and enhance reputation.

By integrating insights from different theoretical perspectives, as well as an understanding of systemic and domestic factors, and applying them to real-world cases, I attempted to contribute to a more nuanced and comprehensive understanding of international relations in the digital age—an area that I find critical and yet unexplored, especially in Latin America.

I propose that this book's findings can be applied to other scenarios, regions, and cases. The increasing competition between the PRC and the US will become more evident in the following years. Technology changes rapidly, but policies and decision-making processes typically have a slower development and transformation rate. Therefore, the interaction between international relations and innovation-related outputs requires constant review but can also be studied in terms of these general rules, international norms, and narratives. Moreover, as new technologies and public processes related to innovation take over Latin America, it will be possible to analyze further if the geopolitical tensions between Washington and Beijing affect specific types of technology transfers and if the region takes a more precise position in terms of securitization, hybrid threats, cybersecurity, digital freedoms, or private data management.

The main limitations I found when reviewing the cases were related to the expected high level of variance in the Latin American region and also the uniqueness of the diffusion of technologies and policies that might seem connected but require very different processes to be deployed, tested, and spread across a country. In this context, the identification of harmonization patterns, as well as middle powers and regional referent states in the technology diffusion process, were necessary to find common points across the different.

Looking ahead, several key trends will likely shape the future of technology in Latin America. These include the increasing importance of cybersecurity, the role of regional and international partnerships in technological development, and the evolving regulatory landscape. As Latin American countries continue developing their technological capabilities, they must balance competing interests and priorities, including economic development, national security, and aligning with global standards.

Ultimately, it is necessary to stress that technology dissemination is not merely a technical or economic issue but a profoundly political one. It reflects broader power dynamics, strategic alliances, and the pursuit of national interests. As technology continues to evolve and shape the international order, scholars and policymakers must adopt a holistic approach considering the multifaceted nature of international relations. Future research can build upon this concern by broadening the number of analyzed countries, the number of technologies reviewed, and the volume of information provided within media inputs. Moreover, comparing countries with different government systems, either centralized or parliamentary, might also offer differences depending on the level of independence regional governments have in the second example.

Notes

1 Boehme-Neßler (2011a).
2 Boehme-Neßler (2011b).
3 Heim and Miller (2020), pp. 6–10.
4 Morgenthau (1948).
5 J. S., personal communication (June 3, 2024).
6 As established in Chapters 2 and 3.
7 Tekir (2020), online; Lee (2020); Tiezzi (2020); Kaska et al. (2019).
8 In fact, some of the cases in this book were considered because of the direct recommendation of experts and policymakers who are working on the issue of technology transfers to Latin America.
9 Morgenthau (1948), p. 4.
10 Keohane and Martin (1995), p. 42.
11 Wendt (1995), pp. 71–72.
12 Copeland (2000), p. 188.
13 Moravcsik (1997), pp. 516–517.
14 Putnam (1988).
15 Nye (2008).
16 Gerschenkron (1962); Gunnarsson (2016).
17 Gilpin (1981).
18 As shown in Chapter 3.
19 U.S. Southern Command (2019); U. S. Mission Chile (2021).
20 Bojalil et al. (2019).
21 Holbraad (1971); Chapnick (1999); Jordaan (2003); Robertson (2017).
22 OAS. (2003).
23 Kobek and Caldera (2016).
24 Anonymous, personal communication (March 8, 2023).
25 E. V., personal communication (August 20, 2024).
26 Tomás (2024).
27 Artaza and Nogales (2021).
28 P. S., personal communication (January 18, 2023); M. P., personal communication (February 13, 2023).
29 Prat (2000).
30 Anonymous, personal communication (March 8, 2023).
31 Dykstra and van der Windt (2004).
32 Portugal-Perez et al. (2009).
33 M. P., personal communication (February 13, 2023); Anonymous, personal communication (March 8, 2023).
34 Carboni and Labate (2018).
35 See Chapter 9 and annex for more details.
36 D. G. and A. O., personal communication (March 8, 2023).
37 Tomás (2024).
38 P. Z., personal communication (June 14, 2024).
39 OAS (2003).

40 As shown in the example of 5G in Colombia. MINTIC (2019).
41 R. M. L., personal communication (March 16, 2023).
42 P. S., personal communication (January 18, 2023).
43 Berwick (2018).
44 Li (2023).
45 Buenos Aires Times (2024).
46 Alter et al. (2024).
47 Leiva (2010).
48 D. G. and A. O., personal communication (March 8, 2023).
49 Hurwitz (2014).

Bibliography

Anonymous. (2023, March 8). *Interview with a Current Public Servant in Chile Who Requested Anonymity.* [In Person].

Artaza, F., & Nogales, D. (2021, November 16). *Gobierno anula licitación de pasaportes a empresa china Aisino tras advertencia de Cancillería y EE.UU.* La Tercera. https://www.latercera.com/pulso/noticia/pasaportes-el-registro-civil-deja-sin-efecto-la-adjudicacion-de-aisino/OZQEJBQUZ5CHVHNAMSJWLXSK5Q/

Alter, J. S., Cook, J. A., & Dussel Peters, E. (2024). Chapter 13. What the ECU-911 Project Has Brought to Ecuador and to China-Ecuador Cooperation. In *Connecting China, Latin America, and the Caribbean: Infrastructure and Everyday Life* (pp. 334–358). University of Pittsburgh Press. https://muse.jhu.edu/pub/49/edited_volume/chapter/3896982

Berwick, A. (2018, November 14). *A New Venezuelan ID, Created with China's ZTE, Tracks Citizen Behavior.* Reuters. https://www.reuters.com/investigates/special-report/venezuela-zte/

Boehme-Neßler, V. (2011a). Caught Between Technophilia and Technophobia: Culture, Technology and the Law. In V. Boehme-Neßler (Ed.), *Pictorial Law: Modern Law and the Power of Pictures* (pp. 1–18). Springer. https://doi.org/10.1007/978-3-642-11889-0_1

Boehme-Neßler, V. (2011b). Cultural Technology and the Law – The Example of Writing. In V. Boehme-Neßler (Ed.), *Pictorial Law: Modern Law and the Power of Pictures* (pp. 19–49). Springer. https://doi.org/10.1007/978-3-642-11889-0_2

Bojalil, P., Egan, M., & Vela-Treviño, C. (2019, February 12). *Data Privacy Reform Gains Momentum in Latin America.* Abierto al Público. https://blogs.iadb.org/conocimiento-abierto/en/data-privacy-reform-gains-momentum-in-latin-america/

Buenos Aires Times. (2024, April 2). *Government Considers Inspection of Chinese Space Station in Neuquén.* Buenos Aires Times. https://www.batimes.com.ar/news/argentina/government-assesses-whether-to-inspect-chinese-base-in-neuquen.phtml

Carboni, O. V., & Labate, C. (2018). América Latina por una red neutral: El principio de neutralidad en Chile y Brasil. *Revista FAMECOS*, *25*(2), 28507. https://doi.org/10.15448/1980-3729.2018.2.28507

Chapnick, A. (1999). The Middle Power. *Canadian Foreign Policy Journal*, *7*(2), 73–82. https://doi.org/10.1080/11926422.1999.9673212

Copeland, D. C. (2000). The Constructivist Challenge to Structural Realism: A Review Essay. *International Security*, *25*(2), 187–212. https://www.jstor.org/stable/2626757

Dykstra, M., & van der Windt, N. (2004). Beauty Contest Design. In M. Janssen (Ed.), *Auctioning Public Assets: Analysis and Alternatives* (pp. 64–79). Cambridge University Press. https://doi.org/10.1017/CBO9780511610844.004

F., R. (2024, June 16). *Interview with a Public Official Serving Argentina's Internal Affairs on Defense Matters.* [Online].

G., D., & O., A. (2023, March 8). *Interview with Authorities at the Telecommunications Agency, Chile.* [In Person].

Gerschenkron, A. (1962). *Economic Backwardness in Historical Perspective*. The Belknap Pr. of Harvard Univ. Pr; WorldCat.

Gilpin, R. (Ed.). (1981). The Nature of International Political Change. In *War and Change in World Politics* (pp. 9–49). Cambridge University Press. https://doi.org/10.1017/CBO9780511664267.003

Gunnarsson, C. (2016). Misinterpreting the East Asian Miracle—A Gerschenkronian Perspective on Substitution and Advantages of Backwardness in the Industrialization of Eastern Asia. In *Diverse Development Paths and Structural Transformation in the Escape from Poverty* (pp. 93–127). Oxford University Press. https://doi.org/10.1093/acprof:oso/9780198737407.003.0005

Heim, J., & Miller, B. (2020). *Measuring Power, Power Cycles, and the Risk of Great-Power War in the 21st Century*. RAND Corporation. https://doi.org/10.7249/RR2989

Holbraad, C. (1971). The Role of Middle Powers. *Cooperation and Conflict*, *6*(1), 77–90. https://doi.org/10.1177/001083677100600108

Hurwitz, R. (2014). The Play of States: Norms and Security in Cyberspace. *American Foreign Policy Interests*, *36*(5), 322–331. https://doi.org/10.1080/10803920.2014.969180

Jordaan, E. (2003). The Concept of a Middle Power in International Relations: Distinguishing Between Emerging and Traditional Middle Powers. *Politikon*, *30*(1), 165–181. https://doi.org/10.1080/0258934032000147282

Kaska, K., Beckvard, H., & Minárik, T. (2019). *Huawei, 5G and China as a Security Threat*. Nato Cooperative Cyber Defence Centre of Excellence, CCDCOE.

Keohane, R. O., & Martin, L. L. (1995). The Promise of Institutionalist Theory. *International Security*, *20*(1), 39–51. https://doi.org/10.2307/2539214

Kobek, L. P., & Caldera, E. (2016). Cyber Security and Habeas Data: The Latin American Response to Information Security and Data Protection. *Oasis*, *24*, 109–128. https://www.redalyc.org/journal/531/53163716007/html/

Lee, N. T. (2020). *Navigating the U.S.-China 5G Competition*. Global China, Brookings.

Leiva, M. T. G. (2010). The Introduction of DTT in Latin America: Politics and Policies. *International Journal of Digital Television*, *1*(3), 327–343. https://doi.org/10.1386/jdtv.1.3.327_1

Li, Y. (2023). *How Are Government's Decisions Affected by an External Power? The Case of Huawei 5G in Brazil*. https://hdl.handle.net/10438/33710

M. L., R. (2023, March 16). *Interview with a Former High-level Politician, Colombia.* [Online].

MINTIC. (2019). *Plan 5G Colombia El Futuro Digital es de Todos*, Ministry of Telecommunications and Information Technology, Colombia.

Moravcsik, A. (1997). Taking Preferences Seriously: A Liberal Theory of International Politics. *International Organization*, *51*(4), 513–553. https://www.jstor.org/stable/2703498

Morgenthau, H. (1948). *Politics among Nations* (1sr. (Digital)). A. A. Knoff.

Nye, J. S. (2008). Public Diplomacy and Soft Power. *The ANNALS of the American Academy of Political and Social Science*, *616*(1), 94–109. https://doi.org/10.1177/0002716207311699

OAS. (2003). *Cybersecurity*. https://www.oas.org/ext/en/security/prog-cyber

P., M. (2023, February 13). *Interview with an Inter-American Development Bank Consultant* [In Person].

Portugal-Perez, A., Reyes, J.-D., & Wilson, J. S. (2009). Beyond the Information Technology Agreement: Harmonization of Standards and Trade in Electronics. *Policy Research Working Papers*. https://doi.org/10.1596/1813-9450-4916

Prat, A. (2000). Spectrum Auctions versus Beauty Contests: Costs and Benefits. *Prepared for the OECD - Working Party on Telecommunications and Information Services Policies*. https://istituti.unicatt.it/economia-impresa-lavoro-OECD-draft.pdf

Putnam, R. D. (1988). Diplomacy and Domestic Politics: The Logic of Two-Level Games. *International Organization, 42*(3), 427–460. https://www.jstor.org/stable/2706785

Robertson, J. (2017). Middle-Power Definitions: Confusion Reigns Supreme. *Australian Journal of International Affairs, 71*(4), 355–370. https://doi.org/10.1080/10357718.2017.1293608

S., J. (2024, June 3). *Telecommunications Consultant with Direct Connections with the International Market* [Online].

S., P. (2023, January 18). *Expert on China's Relations with Latin America* [Online].

Tekir, G. (2020). Huawei, 5G Network and Digital Geopolitics. *International Journal of Politic and Security, 2*, 113–135.

Tiezzi, S. (2020). *China's Bid to Write the Global Rules on Data Security*. The Diplomat. https://thediplomat.com/2020/09/chinas-bid-to-write-the-global-rules-on-data-security/

Tomás, J. P. (2024, February 14). *Costa Rican Court Suspends Exclusion of Huawei as a 5G Provider*. RCR Wireless News. https://www.rcrwireless.com/20240214/5g/costa-rican-court-suspends-exclusion-huawei-5g-provider

U. S. Mission Chile. (2021, July 19). *EE.UU. adopta medidas contra entidades que facilitan abusos de derechos humanos en China*. U.S. Embassy in Chile. https://cl.usembassy.gov/es/estados-unidos-pretende-bloquear-la-vigilancia-del-gobierno-chino/

U.S. Southern Command. (2019). *Adm. Faller Visits Argentina*. https://www.southcom.mil/MEDIA/IMAGERY/igphoto/2002151179/

V., E. (2024, August 20). *International Relations Expert Based in Brazil* [Online].

Wendt, A. (1995). Constructing International Politics. *International Security, 20*(1), 71–81. https://doi.org/10.2307/2539217

Z., P. (2024, June 14). *Intelligence Consultant Focused on Latin American Issues* [Online].

Appendix 1

On Conditions to Analyze Media Outlets for Chapter 9

A total of 44,884 tokens or words were reviewed for Chile on two leading media outlets, EMOL and La Tercera, which have a relevant online presence. In the Colombian case, 42,520 tokens were counted across publications by El Tiempo, which is equally applicable. In Argentina, 29,871 tokens were analyzed across the leading Argentine media, La Nación and El Clarín.

It is essential to establish that the following filters were applied to the new selection and the search. First, news about products, such as cellphones or devices that can use 5G (which is abundant in all countries), was not included as it was mainly used in media publicity or marketing rather than actual news. Examples are the models of Huawei or Nokia phones that support 5G technology, which would have distorted the results.

Second, some of the results could not be evaluated due to paywalls implemented by outlets for certain content. Third, news not produced by the outlet but rather by an agency was excluded from the analysis. In other words, news wired by foreign media and agencies that could have influenced the national analysis of what 5G means for the country was disregarded, and anything written by or based on either Xinhua, AFP, Reuters, the *New York Times*, the BBC, the *Wall Street Journal*, the China Media Group, Nikkei Asia, and so on, along with their narratives, is not considered part of the analysis. All the media included in the analysis had these partnerships to some extent, and it is possible that these international narratives shape local journalists' perceptions of the issue and how local reporters discuss the overall debate.

The program used for the analysis was KH Coder. Using a simplified methodology, it attempted to identify word co-concurrence, keyword identification, and relevance. For each country case and the entire data pool, both a raw, unfiltered analysis (word count) and a specific keyword analysis were conducted. Keywords were selected based on three different criteria: economy, security, and value-related words. In Table A.1, I present general characteristics of the analyzed media and the countries selected for this analysis.

Table A.1 Shows the reach of the selected media outlets that were used for text analysis on the discourse around technology transfers in Latin America, as well as Internet Penetration rates and Level of Freedom under Freedom House's measurements (2023)

Country	*Media outlet*	*Twitter followers of the selected outlet*	*Country's last rec. population*	*Citizen's internet penetration*	*Freedom house classification (2023)*
Colombia	El Tiempo	8.5 MM	52.0 MM	83.2%	70 (Free)
Chile	La Tercera	2.1 MM	19.6 MM	97.2%	95 (Free)
	Emol	2.1 MM			
Argentina	La Nación	3.9 MM	45.8 MM	91.1%	85 (Free)
	El Clarín	3.4 MM			

Table A.1 was created using data from Twitter and the open databases of various media outlets. The population figures were sourced from the United Nations Population Division's World Population Prospects: 2022 Revision, while internet penetration statistics are taken from the world's Internet stats (2022). Emol Chile refers to El Mercurio Online. The Freedom House report is included in the references of this study. The data was last updated on April 14, 2023.

Index

Note: **Bold** page numbers refer to tables; *italic* page numbers refer to figures and page numbers followed by "n" denote endnotes.

For Product Safety Concerns and Information please contact our EU representative GPSR@taylorandfrancis.com Taylor & Francis Verlag GmbH, Kaufingerstraße 24, 80331 München, Germany

Batch number: 10397794

Printed by Printforce, the Netherlands